Make a cow laugh

Also by
John Holgate in Pan Books

On a pig's back

John Holgate

Make a cow laugh

A first year in farming

Pan Books London and Sydney

First published 1977 by Peter Davies Ltd
This edition published 1979 by Pan Books Ltd,
Cavaye Place, London SW10 9PG

ISBN 0 330 25780 3
Set, printed and bound in Great Britain by
Cox & Wyman Ltd, London, Reading and Fakenham

for my wife, Shirley, our family,
and the many friends, biped and quadruped,
who appear in this story

Contents

 1 Off to the country with Old Lil 

March was only a day away when the Holgate family convoy left the tarmacked roads and went plunging and rattling down the pitted, stony lane to Egerton Farm. It was a grey, cold day. The fields were deserted and the buildings empty but the starkness of the scene was redeemed by the mountain which rose, across the valley, austere and beautiful, dominating everything, its white flanks pitted and pockmarked with trees and patches of bracken and gorse, and scarred by criss-crossing tracks.

There was a powdering of snow over the grass, the pond was scummed with ice and there was a bite in the wind but the season was turning. Already there were green knots in the hedges, the black buds of the ash were swelling and in the air there was that spark which says, as clear as someone singing, there must, some time, come an end to winter.

In the forefront of the invasion, driving the family car – an Austin 1800 – came my wife Shirley, petite and lively, currently auburn-haired, with the two 'littles', Victoria Jane, 'Vicky', aged eight, blonde and pigtailed, and Nicholas, four years old, blond, stocky and very voluble. Behind them I steered Old Lil, our newly acquired but aged diesel van, with our elder son, John, sixteen, serious, long arms, long legs, a downy moustache showing on his upper lip, intent on a farming career, in the seat beside me.

Our entrance into our domain went unnoticed by the world around. No brass band struck up, no welcoming committee cheered when we climbed out and stretched our legs after the five-hour drive from London. Nothing, absolutely nothing, stirred. There was something rather intimidating about the stillness. Everyone, even Nicholas Paul, was reluctant to speak.

According to arrangements the key should be hanging on a nail in the outside coalhouse. And there it was with two rusty companions. It turned stiffly in the lock of the scabbed back door and we stepped into the bare, unheated house.

Footsteps echoed and rang as the littles charged about

making discoveries, bagging rooms and shouting to one another. Shirley and I stood in the main living room with its uneven stone walls, low-beamed ceilings and brick floors worn into smooth hollows by the passage of many feet.

She had liked the place on sight but now she said feelingly, 'Well, we have really done it. Just on our own. Just us.'

Her words expressed my own doubts and uncertainties. We were, indeed, on our own. Standing there was very, very different from poring over farming textbooks in suburbia or discussing loans with the bank manager. Our financial resources were very limited. We had stretched ourselves to the absolute limit to buy the farm. There was no one to bail us out if we failed to make a go of it. We had, as Shirley said, 'really done it'.

Exactly why we had done it was something which never ceased to puzzle the locals when, eventually, we met them. They extended their shy welcomes but treated us with polite suspicion. They were right to do so. Why on earth should Jack Holgate, a fat, middle-aged family man, who was doing very nicely thank you, trade the soft, sure comforts of the smoked city for the hard labour of a small farm? Particularly when neither he, nor his wife, knew anything worth while about farming.

Really, the idea had been simmering for years. It waxed and waned. It was something we had often talked about without actually doing anything.

A visit from Shirley's aunt and uncle started it off. They had retired from business and bought a ten-acre holding in Devon: a 'sweet' little house, all modern conveniences, central heating, garage, etc. They considered themselves 'real' farmers because they kept a cow to provide house milk. Her name was Fredrika and she was an aged Jersey. When the uncle milked her, he put on a white overall and white cotton gloves. He had several changes.

'White is more hygienic than coloured,' he said, although why this should be he never explained.

Whenever they arrived at our London home we always asked, 'How is Fredrika?' They always replied, 'Doing very well, thank you.'

A neighbour went in and looked after the cow when they were away.

The rest of their stock consisted of Fredrika's calf, a pig, some chickens, six sheep and about as many cats. They were considering getting rid of the pig because it was too much work. It always needed cleaning.

There were 'farmers' in my family too. Perhaps John's yen for farming was inherited. I had relatives who owned a hill farm in Shropshire where they reared cattle and sheep. They lived in a big, rambling, red-brick house which appeared to be perpetually filled with muddy children, cats and puppies. The sun always seemed to be shining when we visited them. When we commented on the fact, they smiled at one another. Later we realized that 'outsiders' seldom visit farms except when the weather is good.

These characters were always glad to see us. They had few visitors. Our arrival gave them an excuse to stand and talk. The husband was a tall, thin man, overworked, curious about life in London which he had visited once on a day trip, only to get helplessly lost in the traffic. The wife, dark-eyed and apple-cheeked, made us tea in big mugs, apologized for her appearance – she was always busy – and openly envied Shirley's trim figure and London 'country' outfits.

We thought it must be rather pleasant to live in such a sunny, beautiful spot: no commuter trains, no rush, no noise, no cars, and to have friends popping in or coming to spend relaxed, rustic weekends with us.

It was on such flimsy foundations that we built, but somehow our intentions crystallized until, without fully appreciating it, we had decided to 'have a go if the right thing comes up'. I wrote to country auctioneers and was put on their mailing lists for small farms. We studied the advertisements in a farming magazine I bought from a bookstall at London Bridge station.

Suddenly there was Egerton. First an advertisement. It sounded interesting enough to justify an inquiry. The agent sent a brochure. It seemed to be what we had in mind. We decided to go and see for ourselves. We knew that if we were going to try farming, it had to be now. Time was not on our side.

The owners were an elderly couple who were finding the farm's 75 acres too much to cope with by themselves and were unable to find or afford labour. They dreamed of a cottage in a quiet village. They had put the place up for auction but it failed to reach the reserve price. The long, unkempt lane deterred would-be buyers.

They gave us plum cake and tea and the kids played with their cats and a terrier which continually growled but never, we were assured, bit anyone. It was what we wanted but the price was more than we could afford and we went home disappointed.

There was an exchange of letters between us and Egerton and then, quite unexpectedly, they accepted our offer. It was a tremendous shock. We walked round in a daze. It meant putting our house on the market. It sold within a fortnight. I saw the bank manager, talked loans and interest rates, nattered with our solicitor who was extremely enthusiastic about the venture, handed in my resignation at the office, enjoyed a farewell party, and caught the old familiar train home for the last time. It was on time to the minute which, in itself was something of an occasion.

Next morning Shirley and I awoke to the numbing fact that the haggling was over. I had pledged our lifeblood, thrown away our security, and the farm was ours. There was nothing left to do but sally forth and take possession.

In the frantic final days Shirley, always a careful organizer, indulged in a splurge of bulk buying, ticking off items on assorted lists, convinced that we were about to enter the wastelands. The house filled up with goodies, edible and otherwise, until it resembled a wholesale warehouse. A huge deep freezer appeared in the garage.

It seemed advisable to join in while there was still something in the kitty. John and I got heavy working boots – we both took size 8 – from an Army Surplus store near Charing Cross. More important, because we would need transport over and above the car, I bought Old Lil.

Steel-bodied, scruffy blue, with the fading crest of a famous firm on the side, she had been used to deliver wine round the suburbs. She was standing in a Lewisham car lot when I spotted her.

'Perfect for groceries or any other type of goods,' the salesman said when I inquired. 'The side door means . . .'

'Pigs ?' I interjected. 'Would she be any good for pigs ?'

His patter pattern was shattered. 'Pigs ?' he echoed. 'What pigs ?'

I explained about the farm and found him awed and fascinated. Awed that anyone could, of their own free will, actually go and live in the 'sticks' and fascinated at the implications. Not to get him wrong. He loved the country. Had he not, only the previous Sunday, with his wife and the kid – aged six, a little girl – driven into the sticks, right down to Tonbridge ? Had a marvellous day. Called at the local. Played darts. Nice people. No, not to get him wrong, he loved the country. In its proper place. But live there ? He shuddered involuntarily.

We came back to the van. Well, why not pigs ? The previous owners had neatly lined the interior with five-ply. The vehicle might have been intended to carry farm stock. Perfect for pigs.

The engine did not fire first time when he used the starter. It did not even hiccup until he produced a tin of Quick Start and gave it a few whiffs. The vehicle was addicted to the stuff. In all the time we had her, only on one occasion did Old Lil start without a puff of it.

With the aerosol produced, the engine coughed like an asthmatic smoker and stuttered into life. The dealer coaxed it gently into a belching roar and off we went, round the busy streets, wobbling in and out of the traffic, intimidating smart saloons, talking 'pigs'. He knew even less than I did.

Afterwards we sat in his office and agreed a price: £175 to include a new battery, two better tyres and, of course, a full aerosol. An hour later I was driving the van cautiously back to our little corner of suburbia.

Our smart-alec friends spotted her in the drive and congratulated us on finally acquiring a second car. They suggested various modifications to the crest, many of them incorporating a pig's posterior. For our part, we became quite defensive about Old Lil and, it proved, with good reason. The van was to play a vital role in our farming activities.

Her service began that February morning when we took a last look round our house, stepped outside and closed the door behind us. It was cold. The kids were well wrapped in scarves and woollens and there was an efficient heater in the 1800. Old Lil did not believe in such pandering although there was an occasional welcome inflow of hot air from the engine. She drew in a generous shot of Quick Start, shuddered in ecstasy, and swayed forward.

We were on our way! Round the Elephant and Castle, past Parliament, along the Embankment, up over the Hammersmith flyover, looking down on the sleeping city, and on to the M4, heading for Oxford, Worcester and beyond.

It was a journey to a new life, new values, new thinking. We were leaving the concrete city and entering the kingdom of grass. None of us knew just what was waiting or what the future might hold. The only thing certain was that it would be very different from what we had left.

2 A house like a tabby cat

Like many old houses, our new home had an atmosphere. Fortunately it was benign. It needed to be. As we stood there that first day the remoteness of the place began to sink in. From the small, recessed windows there was nothing to be seen but fields and trees against a backdrop provided by the mountain.

It was an old part of the world. Dark-haired Celts had farmed in the area and a Saxon village lay buried and undisturbed under the turf of the top fields. If some of those long-dead farmers could have returned they would have fitted, without too much trouble, into today's pattern of everyday life and work.

The house was in character. It crouched like a grey tabby cat in the middle of a big, neglected garden hedged with sloe and damson bushes, planted with clumps of blackcurrant and roses, Christ-thorn, lilac, apple trees and iris. A golden, cultivated honeysuckle climbed the south-facing front wall and threatened to smother one window: at least it did until the lambs came into the garden, nibbled its tenderest shoots and pruned it far more effectively and ruthlessly than I could have done. A variety of birds, large and small, lived in the roof space, flying in and out of entrances which let in enough draughts to keep a schooner in full sail.

Parts of it were Queen Anne, parts much older, and the kitchen and utility room were a red-brick Victorian extension. In places the base of the rough, grey stone walls was two and a half feet thick. The doors were heavy enough to withstand a siege and furnished with an assortment of latches and bolts. A chimney stack, built on the outside of one end wall, was pierced with small holes put there, so I was informed, 'to catch the wind and help it pull'.

There were five rooms on the ground floor and four bedrooms upstairs with a loo and bathroom where the fifth had been. The walls throughout sloped and leaned as they pleased. Wallpapering was a nightmare. There were odd, inexplicable alcoves. There were recesses cut into the walls for forgotten purposes but ideal for modern ornaments. Where today's fireplace stood in the main room, yesterday there had been a kitchen range with a bread oven above.

One ground floor wall had a large patch that continually glistened with crystallized salt. The best explanation offered was that at some time in the past there had been a pork salting table on the spot.

This was a house that had seen a lot of people come and go. It always seemed to us to be slightly amused at our efforts. It was, Shirley said, 'an intriguing building with great potential'. So much could be done with it. A few days' residence were sufficient to confirm my suspicions that not only was there a lot that could be done, there was also much that needed to be done.

It could have been overwhelming, but our immediate

Predecessors had thoughtfully laid a fire in the grate and a single match set the dry wood crackling and spitting out bright, cheering flames. It stimulated us into action. Like all good Britons faced with such a situation we knew instinctively what to do: make tea!

We had crammed Old Lil with some of the most essential kitchen items. The electricity was on and so, while Shirley brewed up and opened cans, John and I unloaded the van, assembled two do-it-yourself kitchen units with work tops and screwed a couple of wall cupboards into place.

By the time we had finished and had our snack, the removal van bringing the rest of our worldly bits and pieces came into sight, picking its elephantine way down the lane.

Two cheerful Londoners climbed out, slapping their hands for warmth, smug in the knowledge that they had arrived within minutes of the agreed time. The taller man announced, 'Here we are then, Guv'nor, safe and sound,' and stepped through the living room doorway.

There was a distinct 'chunk' as the low lintel bit into his forehead and almost scalped him. He stood, dazed and blinking, eyes streaming, rubbing the spot and repeating, 'Oh my Gawd, oh my Gawd.' Obviously the house had not taken to him.

They knew how to build in those old days. He was the first of many tall men to be humbled. Being myself of a reasonable, sensible height, short to average, it appeared to me that the doorway was divine punishment for anyone presumptuous enough to have more than twenty-nine inches inside leg measurement.

A cup of hot, sweet tea restored him and soon, under the pair's expert handling, the furniture was being moved in, carpets unrolled, beds magicked up the narrow, twisting stairway and all those other items which go to make a home, placed wherever Shirley thought they should 'at least for the time being' be placed.

The removal men, reared in the close hustle and bustle of the capital, were appalled by the isolation of the place. They kept trying to cheer us by recalling similar places they had visited.

'There was a worse place to get to in Devon,' the smaller man,

George, said. 'Right up on the moors. I think it was the edge of Dartmoor. We moved a family from Bexley there. Nice couple too. A bit older than you, I think. They must have been out of their minds.'

His kindly efforts were interrupted by a piercing scream which froze all activity and then sent everyone rushing outside fearful of what might be waiting. We were just in time to see Nicholas, his face red with effort, racing for the safety of the house as fast as his fat little legs could carry him, closely followed up the stockyard by a white, angry, spitting goose. Running third came another, mottled goose, trying hard to catch up but looking very out of breath.

Before anyone could act, Nicholas Paul stumbled and went sprawling in the mud, whereupon the leading bird stopped, turned away and strutted off visibly swelling with pompous satisfaction. I rushed forward intent on murder but the bird simply quickened his pace and disappeared round a corner followed by the other, leaving us to comfort the sobbing four-year-old.

'That bird wanted to eat me,' he wailed.

This was our introduction to Martha, and Moses her villainous husband, two geese bequeathed us by the previous owners. It was a fitting first meeting as we were to find out. Theirs was not a happy marriage, nor were they particularly nice to know.

But at least the incident provided a diversion from unpacking. Nicholas Paul was scrubbed clean and we finished sorting out the most needed items, decided to leave the others until the following day and sat down with the removal men to eat bowls of stew from huge cans, dunking our bread and burning our mouths with mugs of hot tea.

Afterwards we signed the necessary papers, expressed our appreciation in an acceptable form, and said, 'Thank you' and 'Goodbye'. The tall one bowed neatly under the lintel, grinned at us and climbed into the huge pantechnicon to join his mate, and they set off up the drive.

Fifteen minutes later he was back, sweat on his forehead and mud up to his elbows. The vehicle had slid off the icy surface into the ditch and settled there like a stranded whale despite all their wheelspinning efforts to get it back on safe ground.

One cursing hour later, after we had dug and pulled and rammed bits of stone, wood, sacking and whathaveyou under the wheels with no improvement, we plodded a couple of hundred yards back up the lane and followed the fork leading to the farm of our nearest neighbour to seek help.

This gentleman, name Willem, was thickset, with steady blue eyes and the deep tan of an outdoors worker; a kindly, uncomplicated person with the outlook on life of one who expects the worst and rather enjoys it when it happens. He was working in a pen of sheep when we arrived and appeared not at all surprised at our predicament. Instead he nodded without saying anything much, climbed on to a tractor and chugged up to the van. Once there he walked round the trapped vehicle, came to a conclusion, hooked a massive chain to the front axle, climbed back on to the tractor and indicated that the driver should try again. It worked. The tractor's power did the trick. A few minutes later the van was standing on an ice-free section of the drive.

Thereupon our neighbour unhooked his chain, draped it round the front of the tractor and declined a cup of tea.

'It's nothing much,' he assured us in a slow, deep voice. 'Worse things happen down this lane, as you'm bound to find out. What is it you're driving?'

I indicated Old Lil and the 1800.

'H'mm,' he said doubtfully. 'Town cars. This lane'll soon shake the guts out of the likes of them.'

After this cheering prophecy, Willem assured us that if we wanted further help we must not hesitate to call on him. We nodded and our knight errant departed in a cloud of blue diesel-smoke to finish removing clinkers from the tails of his sheep.

The removal men grinned embarrassedly and had another, and this time successful, go at leaving us. We stood and watched as the van, our last link with the world we had left, climbed cautiously up the lane and away. The great body showed intermittently through gaps in the hedges but at last it had gone and the sound of its engine faded.

It was like people on a desert island watching the ship that had marooned them sail away.

No palm trees for us. Egerton, our island, consisted of ten fields of varying sizes set on the steeply sloping side of a valley in a plum shaped grouping with the lane as the stalk and the house, buildings and barn, the pip. Two streams running through steep-sided, wooded gulleys which, in season, were carpeted with bluebells and wild garlic, between them defined most of the farm's limits before merging half a mile further on. The land was blessed with trees: oak and ash, beech, nut and hawthorn.

That first evening, as the light faded, Shirley and I sat tired, more than slightly bemused but content with the day's achievements, in the biggest living-room, not speaking much, just watching the fire burning in the brick fireplace. The logs we had found outside and carried in were slightly damp. They rested on the glowing coal slack and burned fitfully with much crackling and sparking. From the end room came the murmur of the television where John and the kids were happily watching. From outside there was the sound of wind in the big ash tree in the garden and the screech of owls hunting.

Later we left the TV watchers and walked up the lane, perhaps half a mile, to a high point where we could lean on a stile and look out over the saddle of the mountain. It was a cold, frost-clear night with bright moonlight and thin, high cloud. Below us, the nearest some twenty miles away, the farthest over thirty, like illuminations in a distant pleasure ground, a patchwork of twinkling electric bulbs, were the lights of half-a-dozen Midlands towns. Above them the sky was suffused with orange and purple tints. White and yellow ribbons indicated ring roads and motorways.

'This is a different world from that,' Shirley said as if speaking to herself. 'We've stepped into a different world.' She was awed at the realization, very different from her usual bouncy self. It was easy enough for me to appreciate something of how she felt.

The night air grew chilly on our faces and we walked back to the farmhouse. As we did so a huge pale owl materialized out of the darkness and flew silently across our path. It was one of a pair of tawny owls that lived in an old greystone building in a neighbouring field. The kids dubbed them 'ghost birds' but there was nothing frightening about them. They glided over the sleeping fields like benevolent spirits.

3 A bullying gander and self-sufficiency

The following day it quickly became apparent that while we might consider ourselves owners of Egerton, there were other established residents who regarded us as rank outsiders. Most of them preferred to ignore our arrival but some were prepared to contest our claims. Naturally, the latter included the geese.

Poor, fat Martha was harmless enough: a fussy old biddy who wanted nothing but a quiet life. It was not allowed her. Moses was a chauvinist and wife beater. He bullied and chased her. When they were fed, Martha took second choice. When they walked out, she came behind, following her lord and master humbly and abjectly.

When spring unsettled her enough to produce a clutch of eggs, Moses was delighted. He insisted that she sat on them and hardly allowed her to come off to eat, even after the incubation period had long passed and the eggs were obviously addled and not going to produce anything.

Even when the eggs, which were laid in a stone cattle trough in an open cattle pen, mysteriously disappeared, the poor old girl had to continue sitting. There she stayed for another week until we investigated and discovered that she was trying to hatch a round piece of limestone. Martha was not very bright.

For all of us, but especially for Nicholas, until we realized that Moses was a fraud and a bully, walking round the farm proved to be a nerve-racking experience. The gander would wait in hiding and suddenly rush out screeching and hissing, wings flapping, neck outstretched, threatening instant murder. The instinctive reaction was to run. This was just what he wanted. If you did, and some of us did, he came along behind prepared to peck anything within reach. A plump little boy in short trousers offered him plenty of scope.

The answer proved to be a stick. Moses knew well enough what one could do. The children and he were soon engaged in open warfare with no quarter asked or given.

The breakthrough came when a carefully laid trail of bread-crusts led the gander and his Moll into an ambush prepared by Nicholas Paul and his sister. Our four-year-old son bounced up from behind some straw bales and fired his popgun at the horrified geese. It had lost its cork but could propel small stones. Vicky, a pheasant tail feather stuck Indian-style in her fair, braided hair, leapt to the attack shouting, for some strange reason 'You dirty rat, you killed my brother.'

Poor Martha let out a shriek of pure panic and the gander about-turned so quickly he lost his balance and went rolling into the mud to be prodded to his feet and chased down the yard by the two victors. It was a complete rout.

A few days later a visiting neighbour expressed surprise at seeing Moses peer cautiously round a corner before venturing into the open. Never, he said, had he seen such a nervous goose. Nor, he said, had he ever seen such a dirty goose. The explanation for the second condition was that Moses had to cross the stockyard to reach the pond and get cleaned up and this he simply could not bring himself to do. After the incident he contented himself with scaring visitors to Egerton, not us; but even among those he was discriminating and never, never went for small children.

Then there were the mice. They were there in colonies so well established that they scorned the attentions of Fanny Fatcat, our London import. They came and went as they pleased, secure in their routes inside the thick walls – until the visitation.

This was rather eerie. For a week or more we heard mysterious squeaks and frantic patterings and scurryings inside ceilings, behind panelling and skirting boards.

The explanation came when Shirley bolted out of the garage (which was built on to the house) gabbling almost incoherently about a big, red caterpillar which moved like lightning. It was, according to the brief glimpse she stopped to get, about five inches long.

No doubt about it: our visitor had been a stoat. He must have been leaving the premises because there were no more noises from unlikely places. Perhaps he was tired of a diet of house-mice. Whatever the explanation, he had done a splendid job. Our mouse population never recovered its previous numbers or confidence and came well within the cat's capabilities.

It soon became clear that our forefathers must have been hardier folk than we. Perhaps we had been softened by centrally heated homes. Whatever it was, we became aware in the days that followed our arrival that most of the warmth generated by the fires floated upwards and vanished through the ceiling.

I set out to investigate the possibility of insulating the loft. My appearance there, via a rickety stepladder and a tiny trapdoor, triggered off an explosion of birds, mostly sparrows and starlings. No problem about them escaping: there was a hole a yard square in one end wall. And they didn't go far; most of them simply flew around, returning to perch at a safe distance and watch me prowling about.

This was my first encounter with a loft I was to know well: too well. It was frighteningly apparent that we had chronic woodworm and it was so cold that the freezer we had brought with us seemed rather superfluous.

The birds considered it their preserve. When I put down rolls of fibreglass material at a later date, they sat and glared balefully, resenting the intrusion. They didn't like the stuff. It got in their feathers and irritated them. They struck back by keeping us awake as they squeaked and complained and jostled for the perches clear of the stuff. We used to have to wait until they had settled down and gone quiet before we could get to sleep.

The main features of that loft are imprinted for ever on my mind. It was there, on the fifth day of our occupation, that we had our first lesson in rustic self-sufficiency. This was all a matter of pipes.

Egerton was well off for water, having both mains and local supplies with an electric pump feeding the latter to a tank in the loft which, in turn, fed the hot water system and the domestic cold taps. The mains came in via a tap in the kitchen which was marked 'HOT' but in fact provided numbingly cold water.

Unfortunately the electric pump had been switched off by our predecessors and not switched on again by us. I discovered this oversight when the taps stopped running. That was quickly put right but still no water flowed. Mentally donning my plumber's hat, I diagnosed an air-lock. It seemed reasonable. Besides I could not think of any other possible explanation. So I began unscrewing pipes in the kitchen and pushing odd bits of wire about and rattling things. Nothing happened, so we went through the yellow pages and found a plumber.

This gentleman was interested right up to the point when I mentioned the address whereupon, magically, his book was flooded with jobs that had to be done that very day, within the hour, no delaying them. No waiting until tomorrow which was also booked up solid. But he was prepared to agree that it was probably an air-lock and to make helpful suggestions.

'Have you,' he asked, 'tried blowing down the pipes?' Had I not? 'Ah, but in the kitchen. It could be near the cold water tank,' he suggested. 'Have you tried there?' I had not. 'Ah, well . . .'

So up into the loft to unsettle the birds and unscrew pipes and blow. And blow. And blow. My efforts continued for the rest of that day and the best part of the next. They elicited distant murmurings and bubbling sufficient to maintain interest and suggest that success was just a puff away, but they never, never produced the much needed water.

It was achingly cold. The feeling began to leave my face. I descended frequently to rest my lungs, drink hot soup and thaw out my blue lips. At last I conceded defeat: obviously the old bellows were not what they might once have been. I telephoned the plumber and was fortunate enough to catch him at home between urgent jobs.

'Perhaps you haven't blown in the right place,' he said.

It was a possibility.

Having the trapdoor open all day made the house like an Arctic wind tunnel. My family are not very reticent. They went round in sweaters and Balaclava helmets wailing about how cold they were. The two littles kept climbing the stepladder which was unsafe and trying to peer into the loft.

During one of my descents, Shirley, almost lost in my biggest sweater and wearing a grey Balaclava which made her look like a big-eyed baby seal, tried to be helpful.

'Why don't you phone Mr Lane, the little plumber who fixed the tank in the other house?'

'Don't be damned stupid,' I snapped back, more angry at myself for not succeeding than at her. 'He's in London. What the hell can he do for us?'

'He might be able to suggest something,' she said in her best tight-lipped, why-is-he-snarling voice. I said something very rude and immediately felt a heel.

Nicholas came pounding downstairs to interrupt and inform us, unnecessarily since we already knew, that the loo was not flushing.

'Then you'll have to use the outside loo,' I said, feeling the satisfaction of making them suffer with me. It was a brick ice-box.

We all went to bed that second day feeling very sorry for ourselves, with Shirley trying to stop the littles making comparisons with our 'other home'. It had already become known as our 'other home'.

The pantomime was set to begin again the following morning but desperation brought inspiration. I was prepared to try anything rather than face another day in the loft. I coupled the high pressure mains supply to the low pressure local supply, joining the two taps with a length of rubber hose. It did the trick. In a short time the cold water taps in the lower regions of the house were functioning, even though the taps in the bathroom remained dry.

Then, with a flash of genius but without logical reason, I unscrewed the pipe supplying the bathroom loo cistern and blew down it until my ears popped. By this time I was an expert blower. There was an ominous gurgle and a disgusting belch and water exploded into the room. Before I could tighten the connection everything was soaked, my clothes, the walls, towels, curtains, floor. But never mind, the water was running again and we had learned an important lesson in survival. It was that what had previously been done for us, we must now do ourselves: there was simply no one else to do it. It was a lesson that would be repeated again and again.

The cold sore that had blossomed on my bottom lip took over a week to clear.

4 Milking with a kick to it

Our rapid transition from armchair to broken-nailed farmers was necessitated by a limited and diminishing bank balance. The knowledge that there would be no salary cheque at the end of the month sent shivers down my spine. We simply could not afford to remain idle for long – but where to begin?

It was painfully obvious that sitting in suburbia 'talking' farming on the strength of books bought or borrowed was one thing. Doing it was quite another. Neither of us had ever milked a cow in our lives but we were going to have to learn, and fast, because all the authorities we had read agreed that dairying offered the best chance of making a living – and we had to do just that – on a small farm. Our need was for practical experience. Fortunately we found the worthy Griff and his pub.

The Forge and the little auction ring across the road where periodic sales were held throughout the year, was the heart of our new world. Once upon a time the pub really had been a blacksmith's shop. The heavy shires that supplied the power on farms in those days had been brought here to stand stamping and snorting, waiting their turn to be shod. So too had come the cobs and carriage horses, hacks and hunters and, of course, the Roman-nosed Welsh Mountain ponies so loved by children.

In one section of the long, low building the blacksmith had brewed his own special, treacle-black brew. It was potent. Many a farmer had been glad to be helped on to a broad back and have a slap send the horse off to find its own way home.

Those days had gone but the pub was still vital to the community. Its wares satisfied the locals' thirst and washed the dust from their throats. Petrol pumps in the forecourt catered for the energy-hungry horse substitutes. There was a café with a deep freeze filled with convenience foods for busy country wives and ices for their children. There was the public bar where the locals gathered, and a neat saloon bar mainly patronized by the well-dressed townees who came roaring along the winding lanes in their cars with their girls alongside to spend an evening in the country. In terms of miles it was a short journey for them but the mental gulf between the two groups was almost unbridgeable.

It was a quiet pub, even on darts nights. When the weather was 'down' and the mud of field and lane clung to heavy working boots no matter how much you scraped them on the door mat, it seemed sacrilege to walk across the scrubbed, red-tiled floor but Griff and his pert, healthy-cheeked wife never appeared to mind.

He was a plump, fatherly figure in his early fifties with bushy, upswept eyebrows which appeared about to fly away, a natural host and very much a round peg in a round hole. His wife was the ideal partner for him. No matter when you appeared there were always the same ready smiles, a quick rundown on the local news, and the latest market prices. All in the local accent which owed much to the Welsh. If we wanted to contact anyone, find a craftsman, get a job done quickly, borrow something, the Forge was the place to go. The benign Griff was the local Mr Fixit. He knew just about everyone, knew who did what or could get what, and what it ought to cost.

'So you'm the folks that's taken Egerton,' he said the first evening we called in the pub. 'We've all been wondering what you were like.'

He introduced us to the listening locals. Most of them were farmers like Old Jonathon, a wizened, sun-burned gnome of a man who peered at us suspiciously before moving along the bench seat and saying with twinkling eyes, 'Come a bit closer to the fire. You don't look daft, so why have you come a' farming?'

The accepted form of dress seemed to be sports jacket and flannels or old suit trousers and well dubbined boots. But Old Jonathon stood out in a heather-coloured, hairy tweed suit of an early vintage with striped shirt, white collar and string tie. The end result was strangely Edwardian. It was, we eventually discovered, his 'Sunday Best' and was being worn only because he had been into town on business. Normally he, too, conformed with a sports jacket although usually with a brown 'shop assistant's' smock on top.

They were as curious about us as we were about them. London, to Old Jonathon and his even more ancient cronies, was as remote as the dark side of the moon. I have never heard him mention visiting any place much more than thirty miles from where we met. His idea of the capital varied according to what television programme he had last seen. The city was always a concrete anthill teeming with people and lethal motorcars but, enlivened by beautiful, fur-clad ladies and tipsy millionaires or, alternatively, by dangerous criminals robbing banks and having running fights with tough 'coppers' along the busy pavements. 'No, no,' he said when some of the younger, more travelled men tried to put him straight. 'It was on TV last night. I saw it myself.' We preferred his version of London life to the one we had led, so we did nothing to disillusion him.

Eventually we came round to our problems and, specifically, how we were to learn about milking cows. Rather, it was how I was to acquire such knowledge, since Shirley was not yet mentally conditioned to come nearer than absolutely essential to the beasts, let alone touch them.

'Going to milk a bit, are you?' Griff said approvingly. 'It's the right thing, that's for certain. If there's money in anything these days, it's in milk. What do you say, Billy?'

That gentleman – Old Jonathon's employee and a bachelor – raised his eighteen-odd stone from the bar, revealing a hole in the elbow of his jacket, and agreed. 'Yes indeed. A few cows, a few calves, a few sheep, a bit of everything – but no pigs. Don't you have anything to do with pigs. Not unless you needs someone to talk to. That's all there is in them. Nothing but muck and company.'

Someone else joined in. 'I don't know that I could agree with that, Billy. They were going a bit at the market on Monday. Going a fair little bit.'

But the big man was unrepentant. 'Pigs would be all right if they'd no stomachs. They eat too much. Much too much. Eat you out of house and home.'

I bought them halves of bitter beer to top their glasses and we sat and discussed my requirements. It astounded them that anyone could reach middle-age and not know how to milk a cow. They needed convincing that I was not, for some obscure reason, pulling their legs.

'By God,' a red-haired character exclaimed, 'I mind milking a cow when I was not seven or eight years old before I got off to school in the morning. A funny old cow she was too. Kick like a horse. Nothing personal about it, mind, but no use calling to my dad, he'd enough to do on his own.'

That brought the subject round to cows that kicked and off they went, apparently forgetting me and my needs.

'Old George was walking past her one day carrying a full churn, never even touched her, but out she swings, neat as you please, and sends him arse over tip into the muck, milk and all. So up he gets swearing fit to shame the devil and lashes out at the old bitch, only he forgets he's wearing Wellingtons, and bang goes his big toe, clean break. Limped about with it for weeks . . .'

Everyone laughed and another gentleman who looked as if his face had been left outside in the rain too often, said, 'The only cow I was ever properly afraid of was just a wisp of a heifer, a first calver. But kick? Like a horse. Like lightning. And before you'd even laid a finger on her. Catched me on the hip one day. It ached for a week and more. Right in the middle of the hay time too. She was an evil biddy but we come off best in the end . . .'

He paused to sup beer and someone asked, 'How'd you cure her, then?'

'We never did,' he said slyly. 'I never said we did. We sold her, straight as you please, put her into the ring and was open about it all. "She's a kicker, gents," the auctioneer says. "That's why she's here." Then he looks hard at old Reggie Meredith from The Top, like we'd told him to. "Take a better man than you to handle this one, Reggie," he says.'

'What happened?'

'Well, you know old Reggie. He wouldn't have none of that, not with everyone looking at him and laughing. "By God," he says, "no cow's too much for me." "This one is," the auctioneer says. "No, by God, she isn't," Reggie says, and he buys her and takes her home. Paid a fair price too . . .'

Everyone was waiting. 'Did he quieten her?'

'Did he hell,' the craggy one said, savouring the memory. 'Did he hell! Two weeks later he was back at the market walking on a stick. "How's that cow going, Reggie?" my old man asked him, sweet as a nut. "You bloody crooks," he says. "Her's over there in the fatstock, let the butchers deal with her. She's crippled me, crippled poor Tom, near scared the wife to fits . . . and our bloody cat won't come near the milking parlour" . . .'

After the laughter, Griff bustled around filling glasses at a sign from Old Jonathon, and somehow, when it was time to go everything had been arranged. I would go to Ellis to be initiated into the mysteries of milking cows. He was a small, neat man with sharp, bird-like features, the patience acquired in long years of handling dairy cows and a quiet, tolerant humour. Throughout the evening he had listened and laughed without speaking much. Now, about to leave, he put on his overcoat and cloth cap and said simply, 'You'd better start tomorrow afternoon. Be at my place about four o'clock, it's down the road. You've a bit to learn.' He bid Shirley – 'the Missus' – goodnight and left.

'There's no one knows cows better than that man,' Griff said as he dried glasses. 'He'll put you straight.'

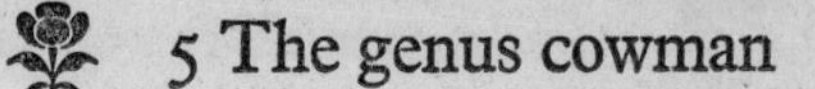

5 The genus cowman

Ellis was my introduction to the genus cowman. In farming they are a race apart. Dedicated professionals with deft hands, an instinctive sympathy with cattle, and the gift of charming a few extra gallons from the herd. They climb out of bed, summer, winter and spring, in those dreadful dawn hours when lesser mortals are still deep in sleep and they deem it an unforgivable sin to be late with their milk and so delay the collection lorry which takes it to the central dairy.

My teacher was already at work among his beloved cows when I arrived for the first lesson. There were two rows of twenty black and white Friesians in the long building. Each cow was tethered in place by a neck chain. Every two cows shared an open cubicle separated from the next pair by tubular steel barricades. In front of each cow there was a concrete feeding trough and a water supply operated by the cow pressing its nose on a metal plate.

A wide, sunken aisle ran the length of the building between the two rows of cattle. It was called, charmingly, the dunging passage. The name was amply explained by the fact that the nether ends of the cows pointed towards it and, in the case of the bigger cows, tended to protrude over it. The arrangement facilitated cleaning operations.

For the first time I fully appreciated that Friesian cows are big animals. They weigh between ten and thirteen hundredweights each. From where I stood they looked positively enormous, elephantine creatures. I experienced a marked sinking sensation. Where, I wondered, were those cows that appear in the telly ads, the ones that look as if they've just stepped out of a bubble bath? And where were those gorgeous, diaphanous-clad creatures who always appear to be drifting languidly among them?

With the best will in the world honest Ellis was no substitute. He was wearing Wellingtons with a rubber apron protecting his clothes. And the cows, kept indoors since the previous

November, had not always remembered the purpose of the dunging passage. Sometimes they had deposited on the cubicle floor and then slumped down happily into it with the result that by now, with spring sighing in the hills, their backends were armour-plated with bubbles of hardened muck. It really was all very different from the television version.

'Be with you in a minute,' Ellis said. He was fitting the milking units on to a cow with hatrack ribs and feet as big as manhole covers. Had I met her in a field I would have run for it.

'Anything I can do?' I asked. It seemed the appropriate thing to say.

He looked curiously at me. 'I thought you didn't know what to do.'

'I don't. But I thought it right to ask.'

Ellis grinned. 'Don't be nervous. Just watch what I do for a bit. You'll soon see what's what.'

There are many milking systems in use but they have the basics in common. Ellis used a bucket system, moving his two milking units down the line from cow to cow, coupling them up to a hissing vacuum line which went round the building above the animals' backs. The vacuum pump and the electric motor that ran it were tucked away in a tiny room in a corner of the big main building.

As the four-gallon stainless steel buckets filled, he emptied them into the ten-gallon churns lined up at one end of the building, sieving the warm, frothing milk through a fine, disposable filter. Finally, the milk was cooled by passing it over a 'washboard' cooler inside which cold water continually ran.

Some of the cows turned their heads to look at the stranger, but most were far more interested in Ellis. They were eager to get the dairy nut concentrates which went with the milking.

In that district, and probably elsewhere, cows were said to be 'paid' for their milk. 'You mustn't steal from them, and you must never cheat,' Ellis told me earnestly. 'They'll know if you've served them short.' I had visions of a cow with a pocket calculator working out that her allowance was half a cup low. He saw my expression; 'You'll see. There's nothing to be gained by cheating.'

At this time of the year, the going rate was high. The cows got four pounds of concentrates for every gallon after the first. They were expected to produce the initial gallon from their maintenance rations. Even to me the fresh dairy nuts smelled good. Judging by the cows' anxiety, the way some of these four-legged milkbars slavered, to them the concentrates must have tasted like a combination of truffles and caviare. It was also soon apparent that although Ellis's precious cows might well be gentle, sociable creatures in most things, dairy nuts brought out the defects in their characters. If there was any chance of getting a few extra, they were prepared to steal, cheat or bully their best friends.

Even Ellis could not deny it. He shook his head sadly. 'You'm right, Jacky, I'm afraid. They'd strangle their grannies if there was dairy nuts to be had.'

It was, according to him, the only flaw in their otherwise perfect characters.

We progressed down the row. Before the units could be fitted, the cows' udders had to be washed clean with warm water to which a few drops of sterilizer had been added. Now this might be an easy enough task in summer when the cows were lying out in the fields, but it was a very different thing in winter when they were kept indoors. Often their bags were fouled with dried muck which called for a pot scourer more than anything else. But patience was advisable. Ellis's paragons did not believe in suffering in silence, they were apt to kick off the units first, trample them into the muck second, and punish the washer third.

Naturally, the possibility of any such thing happening to Ellis was completely unthinkable. For him everything went easily and smoothly. He arrived at a cow, washed and cleaned the udder and fitted the rubber-lined teat cups of the units as naturally as scratching his head. The concentrates rattled into the troughs and if a cow was impatient or stirred he simply said, 'Steady, old girl, steady now,' and everything was well. The pulsators on the units plip-plopped away, the cups moved rhythmically, and the milk began to flow into the bucket.

Nothing, according to him, could possibly be easier. That

surprised me; I could think of quite a few things. After following him round, carrying the wash bucket and sponge for half a dozen cows, I was allowed to wash a couple of the quiet ones. Then came my big moment.

'Her's a dozy old girl,' he said, indicating a gigantic creature which was chewing the cud with a jaw movement that looked well able to cope with a man's arm. 'Try her, she won't do anything daft.' It was me doing something daft that I was worried about. Nevertheless I struggled to put on a brave front.

The washing ritual took twice Ellis's time but I managed it without too much trouble. The cow went on chewing and, encouraged by her disinterest, I ventured to fit the units. They went on easily enough. At least three of the cups did. But on the wrong teats. I was left with one teat cup and no teat within reaching distance.

I took them off and began again, determined not to repeat the mistake. Just to make sure, I bent down and studied the cow's equipment. There were four teats all right: I had half expected to find one missing. Ellis, affecting not to notice, took out a polka-dotted handkerchief and blew his nose.

This time the cups went on the right teats, but the strength of the suction surprised me and they pinched her ladyship. She had previously ignored my fumblings, but now she turned her head and studied me carefully. Then, with great deliberation, she lifted one outsize back foot and swept, rather than kicked me contemptuously out of her sight. I went back several feet before ending up on the concrete, fortunately in a muckfree spot.

'Two things, Jacky boy,' Ellis said, wiping his nose. 'First, you forgot to feed her. Second, never pinch their tits. They don't like that, not one little bit.'

It was back to carrying the bucket for another few cows but then, because there was really no option, I plucked up enough courage to have another go. This time everything went more or less according to plan. It all fitted together with nothing left over.

'Better,' Ellis said approvingly. 'Now you'm getting the hang of it. We shall make a cowman out of you.'

Such praise! I felt like a fourth former for whom the Headmaster has just prophesied a brilliant future.

John, who had been found a place at the grammar school in the nearby town, was envious and would dearly have loved to change places with me. The younger pair had started at the local village school four miles away and considered their own lessons much more interesting.

Towards the end of the first week – I always went in the afternoon because in the mornings with the collection due he was too busy to stop and explain things – he said, 'You ought to start looking round for some cows, Jacky. Get a few in to make a start, then add a few more. They'm selling up at Oxenford the end of next week, whyn't you look for something there?'

'How about you coming with me?' I asked.

'Oh, I'm going,' he said grinning. 'Why d'you think I said it? We could look at the cows and see if there's anything worth spending money on.'

The intervening week was useful. It gave me time to prepare the milking machine which we had purchased with the farm, obtain any extra tackle needed, and complete the necessary formalities.

In our innocence, now rapidly eroding, we had thought you simply arrived at a farm, got yourself a cow or two, somehow extracted the milk from them, and, hey presto, there you were in business. Not so. If you want to produce milk for sale in Britain, you must register with the Milk Marketing Board.

Nor is this a rubber stamp process. It entailed, first, while I was being initiated into the mysteries by Ellis, a visit by a very competent lady from the Ministry of Agriculture who gave our premises a stringent inspection before saying she would recommend our application for registration.

Shirley plied her with tea and cakes to show our gratitude.

She was followed by a very earnest and very adenoidal young man from the Milk Marketing Board who brought the contract for us to sign. He quickly made up his mind that neither of us could possibly understand it. He insisted on going through the document word by word and declaiming what he considered particularly important sections in an atrocious Black Country accent. Every now and again he paused to be sure that we poor yokels fully understood the significance of what was being said. 'Yes, yes,' we would say, trying to give the impression of hanging on every word. 'What comes next?' We would have signed the thing even had it meant us committing hara-kiri should Egerton milk fail to pass the depot tests.

A few days later two iron milk churns appeared, like a magician's props, on the stand built with railway sleepers at the top of our lane. It was from this altar that our offerings would be collected daily, not later than 8 a.m. We also got a supply of labels printed with our name and identification number, with a space in which we should write how many gallons of milk we were sending. These tickets had to be tied to the churns to ensure that no one else benefited from our labours when we did, eventually, acquire a cow.

It was this matter of acquisition that Ellis and I repaired to the Forge to discuss further. As we made our appearance Old Jonathon chortled, 'Here he comes, the Man on the Flying Trapeze. Them old cows still playing football with you?' He was in high spirits.

'You show a little respect,' Ellis warned him, 'Jacky's going after cows now. Before long the milk will be rolling out and the money rolling in. It'll be all posh cars and new suits, just like the rest of us.'

'By God, will it though?' Old Jonathon said. 'Then I'd better buy him a drink while he's still poor enough to talk to the likes of me.'

It developed into a rowdy evening.

6 Buying our first cows

The morning of the Oxenford sale found me full of second thoughts about all this dairy business. It was all very well handling Ellis's milkers with the little man standing by ready to intercede on my behalf. It was quite different contemplating milking strange cows on my own.

'Stage fright,' Shirley said unsympathetically. 'It'll be all right on the night. You'll cope. Just try not to buy any three-legged ones.'

She was developing an execrable sense of humour. Farm life, fresh air, exercise, suited her and she looked very fit. Her latest enthusiasm was knitting. She used finger-thick wool, needles like cricket stumps, and churned out sweaters like the orange-coloured one she was wearing very quickly. How she found the time astonished me.

There was obviously nothing to be gained by delaying. I piled on all the clothes possible under an old duffel coat which had once cost me five shillings at a suburban jumble sale, gave Old Lil her morning fix of Quick Start, and set out.

The first person I saw on arrival was one of our new friends, Howard, a short, stocky, grizzled terrier of a man whose farm, measured as the crow flies, was about three-quarters of a mile from Egerton but a good two miles by road because it meant going round three sides of a square. Fifty-odd years old, ex-wartime infantry sergeant and fiercely patriotic, he was proud of being 'self-made' and having, in his own words, built up his place 'from pennies'. It had not been easy. The years of unrelenting hard work sometimes showed in his high-cheeked face and watchful eyes but nothing subdued him for long. We had met in the Forge – where else? – and got on immediately. He was a blunt character with a great sense of humour, as ready to laugh at himself as at anyone else, and easy to like.

Wrapped in a heavy riding mack, he was pottering about among the numbered rows of tools and implements which had been laid out in the field adjoining the farmhouse and buildings. Six hay forks, their twin prongs buried in the turf, stood in a rough circle.

'Shall we take three each, if we can buy them?' he asked.

That agreed, we went on to talk about other things.

'You'll be after cows, I suppose,' Howard said. 'Want to take a look at them before it starts?'

'Ellis is coming along,' I told him and, as if by magic, that gentleman and his son-in-law – Thomas, a tall young man with a bush of black hair – came out of the car park. Their breath vapourized on the cold air as they approached. There followed the usual exchange of greetings and what passed as pleasantries.

'You bin ill?' Howard demanded. 'You look poorly.'

'Na, just a bit of back trouble. Nowt really.'

The son-in-law snorted. 'You couldn't hardly get out of bed this morning.'

'Old age and a warm bed, nothing to do with your back, mon,' Howard teased.

The little man grimaced. 'We'm getting auld, my boy. All of us getting auld.'

Normal relations established, I suggested that we look at the cows, but it appeared that there was plenty of time for that before the start.

This was my first dispersal sale. The auctioneer's catalogue announced, 'Unreserved Dispersal Sale of the Whole of the Live and Dead Farming Stock by Order of Mr E. Sheldon who is retiring from farming . . .' Later I found that the pattern seldom varied. The sale opened with the 'small farm tools' to give all would-be buyers a chance to warm up and then moved to the larger implements and machinery and finally to the livestock.

Today there were twenty-six small-tool lots: items like axes, bale forks and thin-tined muck forks, a worn, sharp-edged garden spade and wide-bellied shovels for clearing animal pens, tins and boxes and jars filled with screws, nails, staples, nuts, washers and bolts, rods for clearing meadow drains, a two-man rip-saw, ropes and wires, and all the other bits and pieces accumulated in a lifetime of farming.

The seller, Edward Sheldon, came up to talk to Ellis and Howard and was introduced to me. Why was he selling?

'Legs,' he said simply and showed the stick he used to walk. 'Something's come to the bottom of me back. Some mornings I can hardly stand, let alone stoop to milk cows.'

'I'd a bit of it myself this morning,' Ellis told him.

'Then I hope God will be kinder to you than He's been to me,' Sheldon said feelingly. 'It's finished me for work.'

It was a leased farm. Now it was going back into the estate even though Sheldon had a grown son, able and ready to take over. Many big estates were operating the same policy, taking back small farms to form bigger units better suited to modern implements and methods. The time of the small, one-man farming unit was passing away under twentieth-century economic pressures.

'But I can stay on in the house,' Sheldon said. 'They've agreed to that. They'll take the land and the buildings, but me and the Missus will stay on in the house.'

He went off to attend to something and Howard said quietly, 'I could never do that.'

'Do what?'

'Stay on and see someone else working my land.'

Ellis indicated his son-in-law. 'He works ours.'

'That's different.' Howard shook his head. 'He's kith and kin. When I goes I wants my own to follow me, not strangers.'

We went on to walk round the rest of the sale. There were eighty-two lots excluding livestock. The centre-piece was a David Brown 880 tractor. It stood, red and gleaming, in the centre of the field. Our little group joined the people looking at it and I was advised to 'watch out for it'.

More men – there was hardly a woman in sight – were arriving every minute in a variety of cars and vehicles that ranged from an expensive-looking Bentley to disreputable, clapped-out bangers. Not, I was rapidly assured, that the appearance or age of any man's car was a reliable indication of his financial standing. A sale like this was clearly an important social occasion. Farmers within a fifteen-or twenty-mile radius, perhaps more, had come to look and haggle, talk to old friends and even, if the price was right, buy. It kept them in touch with prevailing prices and there was always the chance of a bargain.

They tended to be much of a muchness. They came tall and short, slim or heavily built. But they were clearly men physically hardened by work outside in all weathers; their hands were brown, leathery tools that could be turned to half a hundred different tasks. There was an air of self-reliance common to them all; they were used to taking decisions affecting the life or death of their animals and their own profit or loss. In twos and threes they moved among the lots offered for sale, inspecting and appraising, setting values. One man sat in the bucket seat of a hay mower and tried out the various attachments while his friend inspected the two long, multi-bladed knives that went with it.

'Forty pounds, no more,' the seated one said.

His friend dissented. 'Have it at £45 if you need it.'

'Na,' the first one said. 'It's never worth that much.'

Sheep were to be sold in another quarter of the field. They were penned in groups of six or eight according to age. Among them were sturdy Clun Forest and Kerry Hill ewes, cross-breds and small Welsh hill sheep. Two heavy, black-faced Suffolk rams shared a pen.

Interested farmers leaned over and caught ewes to look at their mouths. If you know what to look for, the teeth tell a lot. Sheep get their first pair of permanent incisors at about a year old. The second pair come some ten or twelve months later and a third pair appear after five more months. By the time it is three years old a ewe has a 'full mouth' with four pairs of these front, grass-biting teeth.

The older ewes attracted Howard. Their teeth were well worn. Some were missing. 'I've seen them with none at all,' he said. 'There's an old biddy on my place without a tooth in her head but every year she brings me a couple of strapping lambs and rears them easy as you please.'

'Sometimes they'll get a tooth worn to a needle point so they can't chew at all,' Thomas said. 'Then you've to get the pliers and pull it out.'

Howard prodded the ewes' backs. 'They'd be worth getting, if they'm cheap. Bring a few lambs.'

A fat, burly man, always known as Aaron 'who-farms-up-the-mountain', chipped in on the conversation. 'Nothing wrong with their udders either, Howard. I fancied them, but if you're bidding I'll stand out if you'll buy me a drink.'

'Stand out like hell,' Howard snorted. 'If you fancied them you'd come in if your own grandmother was bidding, you great crook. Don't you try and get cheap beer out of me.'

Aaron, who was wearing a dilapidated blue serge suit, the jacket fastened by one strained button over his bulging paunch, open-neck shirt and no overcoat in spite of the cold, laughed showing a mouthful of strong, yellowing teeth.

'You'd spare a few coppers for a poor hill farmer, udn't you, boy?'

'Buy your beer out of all them damned subsidies,' Howard retorted. 'Making fortunes up there likely enough.'

The other shook his head. 'Howard, my boy, how can you mouth such things? Barely keeping body and soul together. Barely body and soul . . .'

They were long-standing friends and rivals. Aaron's farm was high enough to qualify for hill subsidies on cattle and sheep, a fact which, to the fat man's delight, needled Howard who was not so fortunate.

We finally reached the cattle. There were over seventy of them including store cattle and calves. But we were after the twenty-two dairy cows – all, except one, Friesians – which awaited inspection in a building very similar to Ellis's. They had been carefully groomed and prepared for the occasion and a generous supply of concentrates over the last few weeks had given their black-and-white coats a healthy sheen. One and all chomped away steadily at the hay spread in their feeding troughs.

The one exception was a doe-eyed Jersey. She was smaller and more delicate than her companions. A lot of local herds carried one or more of the Channel Island breed to increase the overall cream content of the herd's yield and so get a better selling price.

'Leave her stand there,' Ellis cautioned. 'It's milk you needs now. Black-and-white uns are the ones to go for. There's no cows like Friesians. They'll give more milk. They'll bring a better calf for sale and when they'm finished you've a good, big barrel of a cow to sell.' It was sound advice.

My companions busied themselves evaluating the cows. One or two people hung back to hear what Ellis had to say. He was well known locally and his opinion was welcomed. Not all the herd pleased him but he conceded that, taken overall, they were a worthwhile lot.

'Now that there's a proper cow,' he said suddenly, indicating an angular beast, all ribs and hip bones. 'Not a bad quarter in that bag. That's what you've got to look for; a cow that puts the feed into her bag, not on her back. She'd give six gallons on the grass, I'll be bound.'

'Put a price on her,' the practical Howard said.

'Give away at one hundred and sixty,' Ellis's son-in-law, Thomas, said. 'A cow no better than her fetched one ninety at Sollars last week. Woolley-by-the-pool bought her.'

'He always was daft,' Howard said simply, 'Mark her down no more than one fifty-five, Jacky.'

Up the line ahead of us, a man swore and jumped back as a cow kicked at him. 'Damn near reached me,' he told his companion.

Ellis and Thomas bent to look at the cow's bag. 'Warts on her tits. No wonder she kicked.' The little man stroked the cow's flank and put his hand gently on her udder. She stirred but did not kick. 'She'd be a good cow without them warts,' he pronounced. 'But best leave her alone. Her'd likely kick your hands off if you wasn't careful with the units.'

Slowly we moved along. By the time we had finished there were seven marks and prices on my catalogue. If I got them all it meant spending more than a thousand pounds.

Howard saw me totting up the figures and laughed. 'Not much profit there,' he teased. 'You'll have to wait a month or two to make your fortune in farming. Let's leave it and get a cup of tea. I'll beggar the bank manager and pay.'

The tea stall is a must at all farm sales. This one, and the bar that went with it, was in an open-fronted implement bay. The old hayracks and the bits of mouldy horse harness which hung about the walls showed that it had once accommodated horses. Now it sheltered tractors.

We got tea and thick sandwiches. A number of locals, some of whom I knew, others I did not, came up and introduced themselves. Several said, 'You'll be buying cows, I know. Shout up and stand in the open if you see me after anything you want.' Afterwards they wanted to talk about what seemed to them the soft, luxurious life of suburbia. Why I had left was beyond their understanding. 'Saturdays, Sundays off and holidays . . .! My Missus thinks I'm daft to stay in farming but wait until she meets you, Jacky boy . . .!' And off they went, smiling and shaking their heads.

A fat-bellied character signalled the start of selling by clanging a handbell and bellowing a mishmash of sounds which might have meant something to the others but meant nothing at all to me. Near us Mark Boyce, the auctioneer, finished his Scotch and sallied forth at the head of his little team. He was a handsome, middle-aged man, expensive sheepskin coat, flat cap and an aura of money, very popular with the farming community. 'He'll let nothing go cheap without having a proper go at it,' Howard said approvingly.

Boyce's clerk smiled at us, picked up a clip board holding the sales sheets, and followed him.

7 Pig troughs, forks and a tractor

Once the sale began we went through the catalogue at the double. Howard got the hay forks for one pound fifty. 'Cheap enough, really,' was the general opinion.

By the time we reached the tractor I was the possessor of three muck forks, four cast-iron pig-feeding troughs, a big axe accompanied by a sack which contained a pair of hedging gloves, a rickety wheelbarrow and a few other bits and pieces.

The tractor was a David Brown 880, 1963 registration and in impeccable order. As we moved towards it a broad-shouldered young man came up.

'If you've the need for a tractor, Mr Holgate, have a go at this one. It's good right through.' He walked away and I asked, 'Who's that?'

'The son,' Howard said. 'Moving down to Worcester way, someone said. Got a job in a factory.'

Bidding for the machine opened at £200.

'Come along, come along, gentlemen,' the auctioneer called. 'Put some sense into it. Who'll give £225?'

I would, and that caused a stir of interest in the crowd.

A moustachioed man offered £250. I made it £275. A lean young farmer took it to £300.

'Hold off a minute,' Ellis cautioned. 'Let's see who's bidding.'

The moustachioed man made it £310. Someone else went up to £320; it was not the lean young man.

The auctioneer was not enchanted. He looked hard towards me. 'Come along, please. A tractor not as good as this made nearly £500 at Kidderminster last week. Who'll make it £330?' I did. This time the moustachioed man dropped out but someone at the back of the crowd waved a catalogue. 'Forty.' I bid fifty. The opposition made it £360. Howard grabbed my arm and held it. 'For God's sake,' he exploded. 'That's my bloody brother Elwyn, bidding against us. Leave it now. We'll sort out the mess later.'

Boyce waited for me but I remained still. It doesn't pay even to twitch at sales. One flick of an eyelid and you've bought a cow or something. I was petrified.

'You're making a great mistake,' the auctioneer said sadly. 'I sell at £360. Going, going, going, once, twice, three times . . . to Mr Elwyn David. And a gift.'

Howard pushed past to catch his brother. The crowd moved on to the grass mower. After a spate of bidding the man who had been sitting on it made the final offer of £55. His friend said, 'I thought you wouldn't go beyond £40?'

'No use,' the other told him sheepishly. 'I've got to have one and it's cheap enough at this price.'

In quick succession a set of zigzag spike harrows, a two-wheel trailer, a fertilizer spinner, a low loader and a dozen other items were sold. The prices met with general approval.

Howard rejoined us. 'Our kid's so shortsighted he didn't know it was us. He can hardly see his hand in front of his face. He wants ten pounds profit. Is it on?'

It was on. Howard went off again to tell his brother and get the name changed on the sales sheets which were being carried back to the auctioneer's clerk's office by a small boy, presumably the son of one of the men helping with the sale arrangements.

So now I had a tractor and owed £370.

As the sheep were being sold a dealer edged over and introduced himself. 'You ever see something I get, you come and talk if it's what you want,' he said. 'I'll always make a price for you.'

We listened to the bidding on the sheep. 'They'm going well,' he opined.

They were: much too well for Howard. Even the old ewes he'd fancied reached £5. 'No profit in them at that price,' he complained. 'Throwing money away.'

The sale moved on until, finally, it was the dairy cows' turn. They were sold in a ring built with straw bales in the stockyard. The big-shouldered son brought them in and walked them round to be inspected.

'As nice a bunch of cows as you've seen anywhere,' the auctioneer enthused in his role as high priest of the sale. 'You all know the farm and it's an honest man you're buying from. No rubbish here. Not a bad one among them. You've looked them over, so let's have some sensible prices. No daft talk. There's money in milk today.'

The first cow was brought in. 'There you are,' the auctioneer sang out, 'a beautiful animal. Fourth calver. In calf to a Hereford. Full milk. Steady four gallons come rain or shine. Like a household pet. Who'll say £200?'

Not a murmur from his audience.

'Come, come, gentlemen, give me some help. Start me off. Say £150—£140 then.'

Someone said, reluctantly, 'Take £90.'

'It's a joke,' Boyce said feelingly. 'But all right, we'll say £90. Now who'll say £100?' A farmer in an expensive tweed jacket nodded. 'Thank you. Now £110?'

Someone winked. '£120 now. Who'll give £120?' No one would. 'Fives. I'll take fives.' No one wanted to give fives. 'All right then, say ones. We'll do it the hard way.' The bidding moved sluggishly to £117. The cow was sold to the man in the tweed jacket.

'Fair. About right,' a man behind me murmured to himself.

The first mark on my catalogue was against cow No. 5.

'Don't you go mad now,' Ellis said. 'Let someone else put a bottom on her.'

No. 5 walked round the ring and seemed to be staring right at me. She looked enormous.

'There's a magnificent animal,' Boyce declared. 'Now we must be near the £200. Who'll start me off . . . say £170.'

Someone said, very drily, 'Say one ten.'

'Well,' Boyce said with a resigned shrug, 'I suppose it's a start.'

Bidding hung fire at £120. It appeared that no one wanted this fine cow if it cost one penny more. Howard nudged me.

'£125,' I offered. My voice sounded like a squeaky toy.

'Aha,' the auctioneer announced. 'Now the real money is coming in. Fresh blood here, gentlemen. Now look after yourselves.' The crowd laughed.

A man directly across the ring from us lifted his catalogue.

'£130.' Boyce turned. I nodded. '£135.' To the other man. 'Come along sir, a beautiful cow. Don't lose her.' There was no response. I could almost hear my heart thumping.

'You've got her,' Howard said.

'Going once, twice. Last chance . . . I sell to . . .' The other man lifted one finger. £136.' Everyone swivelled towards me. I lifted my finger. It seemed the correct response. '£137.'

It was the winning offer. The other chap gave up. 'Going once, going twice, going for the third and last time. I sell to Holgate, Egerton at £137.'

The hammer came down and the clerk scribbled on his sales sheets.

'Well done,' Ellis said approvingly.

'Now you've got a cow,' Howard said grinning like a schoolboy. 'No more lying in bed in the mornings for you, Jacky boy. No more lazy weekends. Now you'm really farming . . .'

When the sale ended I had bought four cows. They cost £137, £104, £126 and £97. The first three had been marked by

Ellis. The fourth I had bought on impulse because she seemed to be going cheaply. She was ten weeks off calving and not producing milk.

'I suppose she's right enough if you can afford to wait for the milk,' Ellis said. 'But it's milk you needs to bring in money.'

The cow that had won Ellis's particular approval went for £196 to the man in the tweed jacket. 'Some folks has got too much money,' he grunted.

Thomas thought we should arrange transport for the cows back to Egerton and then enjoy a quick beer. I thought he was very sensible.

Just half an hour later it had all been arranged. A red-haired lorry driver would deliver the cows. Thomas could drive the tractor home for me. They went to order the beers and I went along to the auctioneer's office to pay for my purchases.

My bill showed £464 for cows, £360 for tractor, muck forks £3.10, axe 90p, wheelbarrow 45p, box of assorted nuts and bolts 50p, partly used roll of barbed wire £1.30, pig troughs £2.10, and a massive great sledge hammer I could hardly lift, 86p. Grand total £833.21. In addition I owed Howard's brother £10 and Howard himself 75p for the forks.

The ground was too soft to bring the van on to the field so after we had finished our beers we carried my new acquisitions into the car park and loaded them into the van.

The young dealer came over to us. He had bought fourteen laying hens. Seven shillings each. If I would give him a shilling a bird profit, they were mine. It seemed reasonable. I paid over the fourteen shillings and asked where they were.

'In that hut over there,' he said pocketing the money. 'They're right enough at the price.'

It meant going back to the sale office to pay for the birds but there was no problem about the change of ownership. This kind of transaction was common. The clerk simply took the cash and made the necessary entry. But the next stage, catching the birds (they proved to be Black Leghorns), was something of a comedy routine. The hut seemed to be jam-packed with flapping chickens and cursing men. Eventually, however, they were all secured, their legs tied with twine, and they were deposited in the van. There, after an initial burst of indignant fluttering and squawking, they lay quietly, occasionally clucking and grumbling.

An unexpected bonus came in the shape of ten eggs laid in the nesting boxes.

'Take them home for the wife,' my friends urged.

It seemed slightly suspect but they obviously considered it acceptable practice and pointed out that Sheldon had retained a sizeable flock of hens in a pen in the house garden. So, after some lukewarm protests, the eggs were wrapped in an old sack and tucked into a safe corner of the van.

 8 The reluctant milkers 

The van was treated as something akin to Aladdin's cave when I reached Egerton. In quick time the hens were carried into a shed which must have served as a poultry house in the past. Once liberated, they walked round like huffy, I-should-think-so spinsters, tucking ruffled feathers into place, and stretching out their necks. A cast-iron trough went in with them for water, and John was dispatched to neighbour Willem to borrow enough chicken feed to tide us over for a day or two. The eggs were gratefully and unquestioningly received by my wife and Victoria Jane and carried off to the kitchen.

That we were now also the owners of four full-grown cows and a tractor was tremendous news for the children. Everything that happened on the farm in those early days was exciting. They hopped around in circles at the thought of 'real' cows Shirley, like me, had reservations.

'Now,' she told the children, 'we shall see whether Farmer Holgate has learned his lessons.'

Her words touched a responsive chord in John who was having

to settle into a new school with O-levels only months away.

'Perhaps we should ask Ellis for a report on his progress,' he jibed. He was standing by his mother and, with something of a shock, I realized that he was probably taller than me. It seemed he had grown overnight.

'It so happens I was counting on John to help,' I told them.

'How else would the older generation manage?' he grinned.

We had prepared the required apparatus in time. Our milking machine was designed to allow six cows to be accommodated line abreast with three of them being milked simultaneously and directly into ten-gallon churns which were suspended from balances. As the milk ran into the churns it was weighed and shown on the clocks positioned at each milking point: ten pounds to every gallon. Once we gained experience it proved a very efficient and fast system.

But experience at this stage was something we did not have. Together John and I checked and then double-checked everything, switching on the electric motor which powered the vacuum pump, adjusting the suspended churns, fiddling with the clock hands, generally fidgeting about. Everything was operative but I was as nervous as the proverbial kitten. After all this was the moment of truth! The cows were on the way.

There were several hours to wait and it was getting dark before the high-sided cattle lorry appeared at the lane gate. The red-haired driver knew Egerton. He had delivered stock there for previous owners. Now he swung the heavy vehicle round like a family saloon before stopping in position opposite the collecting yard.

'Sorry if you've been waiting,' he said, jumping down from the driving seat. 'It's blue murder. You're the last call on this run but there's another couple of loads to be got out before I finish tonight.'

He pulled down the counter-balanced tailboard which fell into position to form a ramp for the animals. Gates, hinged to the inside of the lorry so that they folded flat against the side when not needed, were swung out and dropped down to protect the ramp sides.

Shirley and the kids had come out to see the new arrivals but there was no immediate sign of the cows. They were as nervous as we were. We all waited. One came to the end of the lorry, saw us, snorted and retreated again to the dim interior. Her elevated position enhanced her size.

'Cor,' Nicholas exclaimed, 'it's a giant cow.'

'It looks very fierce,' Vicky contributed.

I thought so too, but it would never do to admit it. 'It's just an ordinary cow,' I snapped, unnecessarily sharp.

The red-haired driver looked rather thoughtfully at me. 'They're just a bit jumpy,' he said. 'I'll get in behind them and bring them out.'

It took a bit of hustling to get them started, but when they did come it was with a rush that took them down the ramp and straight into the lighted collecting yard. John closed the big double doors behind them and slid home the bolt. First part of Operation Cow was completed.

'Well, that's done,' the driver said. He fumbled in a pocket and produced a bruised delivery note for me to sign, refused the offer of tea or beer on the grounds that there was much yet to do, climbed into his cab and left us.

In the collecting yard the newcomers walked round suspiciously. They were to spend the first night here. There was water and we had fixed hay racks. They sniffed everything. They were wild-eyed and excitable and with good reason. Until now their lives had been spent on the Sheldon farm. They had been born and bred there and brought into the milking herd and its routine. Now they were in an utterly strange world. Their udders were swollen and ached with milk. It was practice to milk dairy cows the night before a sale but miss the morning milking to present their bags filled and impressive. They had never been in a vehicle before. Added to this, now they were confronted by humans they did not know. Some of our friends reckoned it could take a full year for a cow to settle into a new dairy herd and produce her best yields.

As John and I stood watching them, first one and then another came up to stretch out their necks and sniff curiously at us with distended nostrils. But there was no comfort or reassurance for them in our scents.

'I hope you can remember something, if not everything,' my son said. He sounded dubious.

The first need now, at least as far as three of them were concerned, was to milk them. In theory it was simple enough. The cows would walk through the sliding door which connected milking parlour with collecting yard and step into the milking stalls. All that would be required then was for us to switch on the machine, put into practice what Ellis had taught me and in a jiffy the place would be flowing with milk and money. Unfortunately no one had explained all this to the cows.

The snag was that to reach their milking positions they had to climb a ten-inch step. This they simply could not bring themselves to do. No sir! They investigated the machine suspiciously, they touched the step with their noses, they appeared to know what was required of them, but beyond that they simply could not bring themselves to go. So they gave a cow's equivalent of an apologetic shrug and decided to forget all about it. It was a nice idea but not for them.

This was, obviously, not a decision we could accept. John and I smiled at one another and resorted to man's superior brain power. What chance could dull-witted bovines stand? Carefully, cunningly, we laid a trail of dairy nuts across the parlour floor, up the step and into the milking position.

The cows followed the nuts like trained gun dogs until they arrived at the step . . . then they stopped. They stretched out their necks and used their tongues to reach as many nuts as possible without actually climbing the step. The rest they were prepared to write off as unattainable.

It was not what we had planned. So we tried putting concentrates into the feed troughs which could be reached only by entering the stalls. We even let the nuts rattle down the metal chutes into the troughs so that the cows could hear.

The quartet knew exactly what was happening. Their mouths watered. They got as far as putting their forefeet on to the slabs and trying to reach the bait. Surely they must go up now! But no. Not one more yard would they venture. It was clearly a time for brute force.

'Come on,' I said to John and we put our shoulders behind a

cow like second row forwards and heaved. Not one inch. It was like trying to move a wall.

'The school rugby team couldn't move this lot,' he said.

We tried slapping their bottoms, hoping to startle them into taking the last vital step. Once we got them there a restricting chain would prevent them coming out again. The slapping produced a very satisfying sound and hurt my hand but it had no other effect. The cows began to get panicky. Vile language was not successful. It added to their determination not to mount the step.

Shirley arrived and asked, 'Would you like a cup of coffee ?'

I snarled at her. 'Of course we would. We would also like to get this so-and-so, stinking, lousy cow up this step.'

'The coffee, I can provide,' she said primly. 'The cows you'll have to manage yourself.'

She flounced out and we went back to trying to persuade the cows to cooperate.

Nearly one hour later we were tired, hoarse, desperate and still struggling. A car pulled up outside and Ellis and Thomas arrived.

'No, no, Jacky. You mustn't swear, it upsets them,' the little man said.

I described their mothers, fathers and distant forebears in graphic terms but he refused to agree and insisted on considering them tolerably well-bred Friesians. Had I noticed how close their bags must come to the edge of the step ? There was a real danger of them stepping on their own teats. By this time it sounded a marvellous idea to me.

'Patience,' Ellis said. 'That's what is needed here. We'll let them calm down and then try again. You'll see, they'll go up when they see what is needed.'

Nearly two hours later, with moon and stars out and shining brightly, we were still trying to get them up the step. Worried cows tend to have loose bowels. These were no exception. My nice, clean milking parlour was an inch deep in sloppy droppings. We all skidded around like drunks on a skating rink.

Next development was the arrival of neighbour Willem with

a coil of wagon rope over one shoulder. He had an unerring nose for trouble. 'I saw the lights on,' he explained dryly. 'Thought you might be having a a bit of a do. You might need this, them here before you needed it once or twice.'

He had provided the answer. Together we hustled and bribed the cows into getting their forefeet on the slab. One end of the rope was fastened to a strong stanchion, we passed it under their backsides, round another upright to give us something to pull against, and hauled them into position. Once there they moved forward happily, began feeding and acted as if all the previous toil and strife had simply never happened.

Three of the four went up in this position. The fourth, one of the three milkers, sauntered in from the collecting yard, looked at her friends happily chomping away and stepped into position as if she had been doing it for years. Even Ellis said 'Damn you' at that!

An hour later we farmers were all back in the house supping tea. The local trio recalled awkward cows they had dealt with in the past. My baptism of struggle was not unique. As they left they cautioned me with wide grins, 'Don't lie on the tail of your shirt in the morning, Jacky. There might still be a bit of trouble getting them up there.' The ever helpful Willem left his rope. Just in case . . .

Shirley was delighted with the idea of 'real milk'. She hurried across to the dairy to see for herself. Minutes later she reappeared with a bucket brimming with the stuff. 'For the house,' she explained happily. Her raid left precious little in the churns.

It seemed no more than five minutes after going to bed that the alarm clock screamed. I crawled out of a warm, comfortable bed, dressed and opened the back door on to a Christmas card scene. It had snowed in the night! I was tempted to go back upstairs but the milking had to be done.

When I entered the collecting yard all four cows were lying on the straw bedding we had provided. They stood up and stared thoughtfully at me. If cows can grin, those cows were grinning.

A firm, masterful approach seemed to be needed.

'Come on, you lot,' I told them in as confident a tone as I

could muster. 'No more nonsense. There isn't time. Let's get this thing over.'

One of them snorted derisively. Another curled her tail over her back and defecated noisily. I pursued the matter no further but simply pulled aside the sliding door.

There was a dreadful moment as they looked at the open door, at me and at one another. Fortune was kind. One after another they walked into the milking parlour and climbed into position. By eight o'clock, muck-splattered, prickly with cow hairs and smelling of hypochlorite – a sterilizer fluid used in washing the tackle – I was waiting at the stand at the top of the lane.

My contribution to the national milk yield was one churn holding five gallons. In terms of hard cash it represented about one pound sterling. The first money I had earned from farming! I wished that the housewives who would eventually get the milk could know about the effort that had gone into its production.

The collection lorry came up noisily and Jock, the little stocky Scots driver, slung up my churn as if it were empty and handed me a replacement. We had a brief chat and cursed the weather. 'It was cold early on,' he said, 'but it looks as though it might improve.'

I watched him drive away and then turned Old Lil and drove her back down the lane starving for breakfast and feeling disgustingly smug. Now we really were farmers. What could we not tackle now? I began to tot up possible production when we got more cows. That was what was needed: more cows. I resolved to get them as soon as the bank manager agreed. Riches, well at any rate solvency, was just around the corner.

 9 A visit to a cattle dealer 

A fortnight and twenty-eight milkings later, I travelled with Ellis to buy another six cows. This time they came from a dealer who specialized in buying dairy cows 'up north' and bringing them south.

He was a short, prosperous gentleman with highly polished leather gaiters who many years previously had put a dairy cow in an auction ring and had been so upset at the distress the noise and commotion caused her, he had refused to do it again. Instead would-be buyers travelled to his farm to look over his stock.

It appeared to have paid him very well. The spreading, red-bricked farm buildings dwarfed our modest homestead. Like many of the 'plusher' places, the farm had been built up in the nineteenth century when there was 'money' in the industry and labour was cheap enough to make building and running big houses possible.

Ellis and the dealer greeted one another with all the ritual proper to members of the same caste.

'How's it going then?'

'Not too bad. Not bad at all. How's trade?'

'Well enough, I suppose. Mustn't grumble. There's one or two fairish cows inside today. Come along and look at them.'

Our host led us into the cowhouse where sixty cows were tied. They were all Fresh Calvers which meant that they had produced a calf within the last week or ten days and were flush with milk.

'No rubbish here,' Ellis announced.

Two other farmers were painstakingly examining the stock, discussing the merits and demerits of various cows and making notes on a pad. They represented competition. We walked quickly along the rows for a preliminary look. Ellis pointed out two cows without pausing to examine them.

'Take that un and that un, Jacky.'

They were £120 each.

Our host grinned and noted them down in his sales book. A minute later the two careful buyers came up to him and began pointing to cows they wanted. He smiled and shook his head: the first two cows they indicated were the ones Ellis had chosen.

One of the two quickly chosen cows was a large, all white animal inevitably named 'Whitey' when the kids saw her. In the trade, so I was informed, such an animal was known as a 'Blue' Friesian. She proved a phlegmatic old girl, somewhat long in the tooth, but within days she was putting six gallons of milk into the churn.

The second quick selection was a younger, handsome, black and white cow with a suspicion of Shorthorn about her. She had a gigantic bag with teats so widespread I had to adapt the units to milk her more easily. But again she was a great producer and added something like five gallons to our daily output.

In view of her ample equipment the bulk of the family favoured naming her Titty. This wakened an unsuspected puritanical streak in Shirley who objected on the grounds that it was unfair to saddle any female, even a four-legged one, with such an embarrassing name. So we compromised: the cow was dubbed Nefertiti and called Titi for short. Shirley was happy with that, she considered it had a continental ring.

'Women . . .!' said John wonderingly.

Ellis and I bought four other cows from the dealer. Two of them alike enough to be twins were named Twin and Sister. A third, boney-hipped cow with a thin, aristocratic face became our beloved Arabella. She was the most intelligent cow we had and the most curious. Nothing happened on our farm without an investigation from this lady. Given the chance she would go walks with the kids, following them happily like some big black and white dog.

The last, a quiet, elderly cow with a bad limp, thought to have been caused by a difficult calving, was taken on trial.

By the time we left the dealer's place I had spent a total of £708. We pulled in to a village pub on the way home to enjoy a quiet ale.

'To wash down the cost,' Ellis said.

We toasted bank managers everywhere. Most understanding of men.

The six cows were delivered at Egerton the following afternoon and steered into the big, covered cattle yard which now housed our other four cows. There was more than sufficient room for them all but they were not prepared to settle for an amicable introduction.

For the next couple of hours we witnessed a dozen bovine contests. There were some Herculean trials of strength. The contestants snorted and pawed the ground and faced one another with lowered heads. Although they had all been polled the thick, horn bosses remained. The object of the exercise was for a cow to push her opponent backwards – something like Sumi wrestling – until defeat was conceded. In the process they skidded about puffing and panting, bumping into their fellows who stood round and watched with wide eyes like spectators at a street brawl. None of the contests lasted long. One cow would give way and, depending on the heat engendered in the struggle, either walk quietly off retaining some dignity or be chased ignominiously round the yard.

Overall victory went to one of the original four, who henceforth was known as 'Gaffer'. She had a secret weapon. Her dehorning had been botched and she had been left with a stump of horn which dug painfully into rivals and soon weakened their resolve.

Not all the cows got involved. The sophisticated Arabella apparently saw little point in it all. Big Whitey was content to let anyone call themselves 'Boss' providing they left her in peace to eat anything and everything edible within her very considerable reach.

By the time they had all settled down and the shoving routine was over, a definite pecking order had been established. Certain cows assumed and were conceded certain privileges. They came in first to be milked and so had first chomp of the coveted dairy nuts. They had first pick of the hay in the feed racks and they took the choicest sleeping spots in the yard.

John and I held our breath that evening when milking time came round. A repetition of the first night with the original quartet was something we could manage without. But when we slid aside the door to allow them into the milking parlour, Gaffer led them in and one after another the first six climbed into position. There were no snags, although Arabella did stop chewing long enough at one point to look round and see exactly what we amateurs were trying to do when she was milked.

The following morning, despite the depredations of Shirley, we put out twenty-seven gallons. At collection time Jock was much impressed. We were doing very well, he said.

As we acquired more cows we found that the step presented little difficulty provided the newcomers were mixed with the established herd and given an opportunity to see how it was done. Even though at times it was nerve-racking to see just how near a cow's splayed hoof came to the precious teats, we never had a casualty caused this way.

There was one sad note. The gentle old cow with the limp just could not cope with the step. I gave her a week's trial but her difficulties were increasing and I telephoned the dealer and asked for a replacement. There was no argument. He guaranteed all his sales. The following day a lorry arrived, took away Hoppity Jane and delivered an all black cow.

The new arrival was an easy-going animal and settled into the herd without trouble, but not all the locals took to her. One visiting farmer's wife, a sturdy, no-nonsense country woman, saw her coming in from the fields and shuddered. 'We'd never have a black cow like that amongst ours. There's something wrong about a cow that's all black. You can't trust them. They can easily turn evil.'

What all this was about and how it began I never discovered. It was a fairly widely held view but one that libelled poor Blackie who was patience and good nature on four legs. But then we discovered there was a lot about local lore we simply did not understand and our friends could not explain. 'It's always been done,' they would say in reply to our inquiries, or, 'Everybody knows that, Jacky, everybody knows that.'

They thought us very curious, but we had no monopoly in this respect. Folk in the area were shamelessly inquisitive about each other's business. We were a favourite topic of conversation and our progress was a subject of great interest. The labels on our churns were carefully monitored. When the milk lorry pulled in at farms, Jock would be asked, 'How's they getting on down there?' There was no chance of stretching the figures when you could arrive at a market or call in the pub and be greeted with, 'I noticed you'm putting out a bit less these days, Jacky. Nothing wrong, I hope?'

10 Frankenstein hands and Arabella injured

Next to our fractured bank balance, the condition of my hands was the best indication of our conversion to farming. The cold weather chapped and reddened them. My city-soft palms blistered and were sore and putting my hands in hot water was a yelping affair until I graduated to callouses. Then in the first few weeks I seemed cursed with the clumsy bug. Every time I swung a hammer it ended up on a finger and left me with a bruised nail.

Seeing my hand resting on the kitchen table one day, Shirley informed me, 'It looks like a hairy crab.' It did too.

It never ceased to intrigue me that while I went round looking increasingly like a worn and tattered tramp, no matter how hard she worked or what she did, Shirley always managed to give the impression of having just returned from some suburban coffee morning. Her one luxury was rubber gloves. She seemed to have brought a five-year supply with her. When I made some remark about expense, she simply pointed to my battered possessions and asked, 'Would you really like mine to look like that?'

That I certainly did not want but it would have been appreciated if she had been able to resist demanding that my hands be produced for our visitors' inspection as if they had come from the Chamber of Horrors. It got so that I tried not to draw attention to them, sometimes even sitting on them.

Being pitchforked into manual labour did more than Frankenstein my hands. My suburban diet sheet soon went on the fire. It was a fight to retain weight, not lose it. Great mounds of potatoes and vegetables, Henry VIII portions of meat, cakes and suet puddings were the order of the day.

For the first few weeks I was so tired it was a struggle to get through the work but it passed. I discovered muscles I had never even suspected and, because of the aches, might have been happy to have left idle.

Weatherwise, late March did nothing to ease my introduction to the dairy routine. The month ended on a bitter note. Each morning for one long, seemingly unending spell, I stepped out into the grey dawn light and was greeted by wind-driven sleet which whipped almost horizontally over the fields and slapped viciously into my face. The only defence was to load up with clothes. Shirley knitted a Balaclava helmet which covered my ears and left only the minimum of flesh exposed. I looked like some polar explorer.

On such days the cows were reluctant to leave their sheltered sleeping spots in the big yard where they had been moved. They had to be prodded to their feet and bullied into venturing the twenty-odd yards to the collecting yard. They scurried along, backs arched, tails tucked in, like old ladies running for cover on a windy day.

The unheated milking parlour was like a cold store until the first half-dozen cows were in position. It soon warmed up: they gave off heat like leather-covered radiators.

But there were better days when the mornings were tranquil. Then I stepped out into a dreaming, drowsy world. Across the valley, higher up the mountain, there would be a spike of light. It came from a door left open. I looked for it. Another man was working there, wrapped like me in heavy clothing, preparing his tackle with numbed fingers, warming his hands on his cows. His name was Mervyn; a small, dark man. We met occasionally and shared the camaraderie of men who earn their living in the same way. We would nod and smile and ask without expecting an answer, 'How's the milking going, then?'

One day in the pub after a sale, he came up and asked, 'What in glory happened at your place the other night? The light was burning until near to midnight.'

It had been a faulty light switch. Rain had got into it. Touch it and you got a very unpleasant shock. So we left it on until it had dried out.

'I was feared you were having trouble with a cow,' he said. 'It sometimes happens.'

He was glad it had been nothing so serious and offered to buy me a beer, but someone else had that in hand and he rejoined his own cronies round the darts board. It was pleasant to learn that he looked for my light just as I looked for his.

No trouble on that occasion but a few days later it was a very different matter. I had the real thing. Fortunately it came at the evening milking when there was more time to cope. It began when I reached under Arabella to wash her udder and found blood on my hand. My heart yo-yoed a few times and sank into my rubber boots. One of her teats had been trodden on in the yard and was very badly gashed.

Now I knew, because it said in one of our books, that a cow's bag is compartmentalized. It is divided into four quarters, all separate from the others, and each, conveniently, fitted with a teat. So I could, and did, block off one teat cup on the units and milk the three unaffected quarters but it was only delaying tactics; the injured quarter could not be ignored or the cow could end up with mastitis.

A nasty complication was that the injury was situated exactly where the lip of the teat cup must come into contact with it. It was going to hurt our Miss Arabella and trouble there would certainly be.

I could not risk getting blood into the milk already in the churn so that had to be taken off and an empty churn put into place. First step was to wash the damaged teat as gently as possible and smear on antiseptic udder cream. The cow was as nervous as I was but she allowed the units to be put on. But the gripping, pulsating milking motion was too sore to bear so she kicked off the units and, for good measure, trampled them into the muck.

This meant that they had to be washed and sterilized before I could try again. Her ladyship was ready and waiting when I returned from the dairy. Her first kick sent the units flying out of my hand. The second kick was intended to send me flying after them. Fortunately she was off target; it only scraped my hip without doing serious damage. But from then on battle was joined. Every time I approached she lashed out.

It seemed prudent to let her simmer down so I milked the rest of the cows and sent them on their way. It was only postponing matters. When I got back, it was to be kicked on the hands, very painfully, and have this followed by another bash on the knee. Finally, in desperation, I called Ellis, my tutor. He was having his supper but swallowed it and came post haste.

'Not one finger have I laid on her,' I said as soon as he arrived. 'Nor have I beaten her, set the dog on her or sworn. But by all that's holy she's tried my patience, the bloody cow!'

He was shocked. 'Don't you say such things and just where she can hear you. You never got such things from me.'

'No,' I agreed. 'I'm sorry, Ellis. She's been a bit awkward.'

He leaned his forehead against the cow's flank and spoke too her. 'Come on now, my beauty. Quiet now. Let's see what he's been doing to you. Quietly now.'

I waited hopefully for that back foot to lash out. But no. Miss Arabella stood there unprotestingly and let him look at the teat.

'There,' he said straightening up, 'we'll soon have that right.' He was speaking to the cow. Before saying anything to me he walked away a few yards and then spoke in a low, subdued voice.

'It's nasty, Jacky, but it could be a lot worse. You'm lucky. If the milk vein had been broken, you'd have to use a cowman's stick. That's something like a kid's sucking straw. You've to put it right into the teat to make a way for the milk.'

His next request was for a bucket of clean, warm water and when this was produced he went to work. Just what he did differently from me, I could not see but fifteen minutes later Miss Arabella had been milked and sent off to join her hay-chomping companions with the cut daubed with udder cream.

'They've got to know you'm helping them, that's all,' he explained.

Frankly, I chickened out of a confrontation the next morning. Instead I milked the others and got my milk to the top of the lane before coming back and tackling the injured teat. She kicked the units off just once but then, heavily bribed with dairy nuts, agreed to be milked. It was a tremendous relief when the milk stopped running and I could let her go.

It was nearly three weeks before her teat healed. I treated her throughout as gently as a time bomb with an uncertain fuse. She fizzed once or twice but never again exploded, mainly because she was too busy guzzling the dairy nuts which I allowed her ad lib. If it was blackmail, I was ready to pay her price.

11 A mob of bovine delinquents

It had always been our intention to rear calves, but nothing heard or read prepared us for the bunch of boisterous, bovine delinquents we acquired.

Local farmers bought in young calves, raised them to yearlings, about five hundredweight, and sold them to other men with lusher lands able to finish them for the fatstock market. There was a keen demand for the little animals, mostly the product of dairy herds, offered for sale in the local auction rings. The calf dealers, like our visitor, were important cogs in the farming machine.

The most favoured rearing method round us was multiple suckling, with several calves being fed by one foster cow. Most of the cows used were Friesian or Friesian crosses but they were very different animals from their cosseted sisters in the dairy herds. In the main they seemed stoical creatures resigned to their role in life, which was just as well because in one lactation of 305 days they could be 'milk mother' to eight, nine, even ten, calves. Somehow, they reminded me of elderly ladies who had slipped down the social ladder.

Our interest brought a prompt invitation from Howard, our ex-sergeant friend, who used the system. He had driven down to Egerton on some pretext, really for a chat and to see how we were managing.

'Come and see for yourselves,' he said. 'Bring the kids and have a cup of tea and a bite. Mind you, like as not, I've been doing it all wrong these many years.'

When we arrived two days later he was just about to begin the feeding. There were four cows tethered firmly in position and the first batch of calves was waiting in an adjoining pen. 'Now you watch these heathens,' he said, opening a door.

Eight calves came in like sprinters out of starting blocks and rushed for the cows. No need to tell them what to do, they each grabbed a teat and began sucking furiously.

'They've got no manners,' Howard said, looking at his watch. 'You've got to see that they don't overload their stomachs. That causes all sorts of trouble. A few minutes and then it's back to the pen. They know the drill, that's why they go at it so hard.'

One cow lifted a foot as if to kick but did not carry out the threat. 'Sore teat,' he said in answer to my raised eyebrow. 'Nothing to be done except rub on a bit of udder cream. She's just got to bear it, I'm afraid.'

With the first batch finished and hustled away, he said, 'Now I'll show you something. Just keep an eye on that old cow at the end.'

The animal referred to twisted anxiously as the new lot of calves came in. 'Looking for her own,' he explained. 'They'll always do it, no matter how many others is tearing at them.'

One of the incoming calves went and stood where the cow could nuzzle him. Not so the others. For them she was simply a milk supply. Howard grabbed the cow's calf and pushed it towards her udder. 'There'd be nothing left if I didn't take a hand. But just keep watching until they'm finished.'

When their time was up, the other calves allowed themselves

to be ushered back to their pen but the cow's own calf stole forward to stand by her head. Our host tried to grab it but the little animal evaded him and got on the far side of the cow which lunged determinedly at her owner, only to be pulled up short by the tether. 'Her'd have me right enough if it wasn't for the chain,' he said and took the calf away, grasping it by an ear and its tail.

From Howard we got the telephone number of a dealer. I rang up, spoke to a woman and told her what we wanted; she promised to pass on the message.

Multiple suckling was not for us. Instead we proposed to try bucket-feeding, which was exactly what it sounded like. Because we needed all the money we could get from selling our own cows' milk, we decided to use a milk-powder substitute delivered by the local Farmers' Co-operative which was made up with warm water. We also obtained six wide, shallow calf-feeding buckets and cleaned out and strawed two pens.

The Monday after my telephone call the dealer's green van came down the lane and pulled up in the stockyard. He was a fair-haired character with an open-air face and shrewd, watchful eyes. His name, we learned, was Roger.

'You'm looking for calves, they tell me,' he said.

He had, I assured him, been truthfully informed.

'There's nine good strong black heifers in there,' he said, indicating the van. 'Not a bad one among them. If you'd the wish, they'm all yours.'

On cue one of the calves bawled.

Farmers when buying try to conceal enthusiasm because it has the effect of increasing prices. With us it was all over the moment the van door was opened. Shirley and the kids were enchanted by the Bambi-like creatures revealed. They stood on stilt legs, stretched out long, glossy necks and bellowed their protests at the world in general. They were babies between a week and a fortnight old. They were tired. They were hungry. And, I half-suspect, they knew suckers when they saw them.

It had been a nightmarish day for them. By van to market, penned with a score of others for buyers to handle, prod and inspect, hustled into an auction ring which must have seemed a bedlam of noise and people, then into the dealer's van for the journey to the farm. Refusing them was never a possibility. I would have been lynched.

'How much?' I asked.

They were on offer at something just over £15 each. Roger, the dealer, reckoned to make about £1 profit on each calf he bought. In view of the work involved, the long day, the transporting and the risks, it was a very reasonable margin. With the littles already arguing over which calves they would have as personal pets, he could have named any price.

The next thing was to unload them. There was a cold wind coming in from Wales and blowing round the mountain. It was, in local parlance, 'very raw'. So we decided to put them all together in the biggest of the prepared pens to keep them warm. It was a tactical error. We should have kept them in smaller, more manageable groups.

Roger, John and I carried them from the van to the pen. They were heavier than I expected. I still had commuter-train muscles. There is not much demand in suburbia for an ability to carry calves. The kids danced about, getting in the way and being shouted at, trying to pet calves and put their fingers into the sucky mouths. But at last they were all in and the door bolted behind them. Four of us went into the house to drink tea, chat and settle accounts. Not so Vicky and Nicholas, they stayed behind to fuss over our spindle-legged acquisitions. For both the farm was a 75-acre adventure playground. They lived in a world separate from ours. Both were happy extroverts eager to be involved in anything and everything. Whereas we had feared they might be lonely, there were not enough hours in the day to suit them.

We offered our visitor beer. He preferred tea and gratefully accepted hunks of fruit cake. It had been a long day for him too, with only a snatched snack in the auction cafeteria. We were his last port of call and he was very ready to chat.

His love was boxing and Muhammed Ali was his hero. Later we found that among the locals he had a reputation for being 'handy with his dukes'.

Eventually we came round to cash. I made out a cheque for £137 and said we would be happy to take another batch in, say, a fortnight's time. He folded the cheque, tucked it into a pocket, and announced, 'Now I'd better give you a bit of luck.' This took the form of several silver coins which he spat on, rubbed against his sleeve and offered.

'What's it for?' we asked.

'For luck. Just for luck,' he explained, embarrassed by our ignorance. 'If you wants any luck with them calves, you'd better take it.'

We took it.

It proved that luck was a widespread custom which sometimes came near to extortion. It was more than commission. There was the same element of forgotten, ancient ritual which ran like a fine thread through so many practices in the area. It would have been more than rude to have refused. It would have been 'unlucky'. Perhaps some forgotten country godling could have provided an explanation.

The exchange completed the transaction. The dealer left and we set about tending the newcomers. Their immediate need was food. Initially they would get three pints of milk or milk substitute twice a day.

Loaded with our six new buckets containing the feed, we set off happily for the pen. Our arrival was greeted by a chorus of bellows which should have warned us, but we were green. In all innocence we opened the door and entered.

It was like the charge of the Light Brigade. Young they might be but they knew that people and buckets meant food and they were desperately hungry. It was a madhouse of heads and tails and legs, splashing milk and swinging buckets. As you tried to feed one calf, another ruthlessly butted its way in, convinced that no stomach in the world could possibly be quite as empty as its own.

Nor was it simply a case of sorting them out and offering them the milk. They might be ravenous and they might know what was in the bucket but they did not know how to get it. They were willing and able to suck but they had to be taught how to drink. This meant putting two fingers into a calf's mouth and leading it down to the milk. By sucking the fingers they took in milk but it was a slow, tedious business.

Some of them learned comparatively easily. Some just didn't have the faintest idea. But the more we fed them, the simpler operations became as satisfied calves withdrew from the scrum and settled down in the straw to rest. It was a wonderful relief when they were all lying down together in one milky-mouthed bunch, all touching, tired and sleepy and contented – if only for the present.

As for us, Shirley, John and I, we tottered out of the pen and took a good look at ourselves. Our clothes were soaked with spilled milk. Our backs ached. Our arms threatened to drop off.

'Ye cripes,' Shirley said faintly, 'do we have to go through that twice a day?'

'We'll split them tomorrow,' I told her. 'Put half in the other pen. That should help.'

'That is a very good idea,' my wife said. 'What about a nice cup of coffee?' If the world was coming to an end she would make coffee.

Nothing less than a stiff gin and tonic was thinkable. We sat by the fire with our glasses not speaking until she said, 'Suburbia was never like this. I wonder if they've filled your job yet.'

12 Alice Capone and other calves

Until the mob entered our lives all calves looked much the same. Not that we had ever given the matter much thought. In reality they proved to have very distinct personalities. The dominant character was a toughie who ruthlessly forced her way to the front at every feeding session. The kids named her Alice Capone. What else?

Our finger-sucking technique was brutally hard on the back.

It was a great relief when, one after another, the calves learned to drink properly. Naturally, Alice was the first to drink. Just as naturally, she determined to use her newly acquired skill to help her slower companions. This she tried to do by wedging her head into the feed bucket alongside the other calf so that neither of them could get anything. When the hungry ones got stroppy and objected, she butted them belligerently and tried to drive them away. A ploy which was especially unpleasant was lowering her head under a bucket when a calf was being fed, and then sending milk and everything flying with a quick, upward butt. Fortunately, she learned equally quickly that certain loud words or a lifted foot constituted a threat, but even so it was unnerving to see her prowling around like a marauding shark as you tried to feed one of the other calves.

Much to the amusement of our neighbours, the kids and Shirley named all the calves. There was Blackeye, a stocky calf with a patch over one eye, and her particular friend, Beauty. They always lay together in the pen. There was White Star and Little Star, String-round-Neck who had a collar of binder-twine when she arrived, and Humpyback who looked miserable and hunched her back unhappily if it was cold. Shirley's particular favourite was known as Mother's Pet, and among others there was poor Plain Jane, the calf who died.

The farm was Shangri-la for the kids. This first batch of calves was nothing less than nine new playmates. Before leaving the farm for the village school (four miles away) each morning they ran down to the pens to check on their current favourite, and they rushed home in the afternoon to find what had happened while they were away.

The calves' response was almost as enthusiastic. They loved being petted and would stand happily as the kids stroked them or scratched them behind their ears. As they grew they played at butting and one day when they were capering round the pens, four-year-old Nicholas Paul announced, 'They're dancing because they're happy.'

It was also Nicholas Paul who dubbed Shirley 'The calves' Mummy', but she objected strongly and forcibly to the inference. She had made herself responsible for feeding the calves and they were quick to realize it. They had only to hear her footsteps in the stockyard or hear her voice to set up a clamour for her attention.

Locally everyone agreed that the danger period for calves was the first few weeks of their lives. 'If you can get them through the first month, the worst is over,' they told us. 'If you can wean them, you'm laughing. The work's all but done, all you've got to do then is stand back and let them grow.'

Just in case we should get too optimistic, they warned us, 'Look out for the scours, especially the white scours. Once they get the white scours, there's not much to be done except dig a hole.' We went round in dread of something going wrong and studied the calves' motions for traces of the dreaded diarrhoea with all the intensity of pagan priests studying oracular entrails.

The fact is that a sizeable percentage of calves die before they are six months old. Most deaths occur in the first few weeks and just as our neighbours warned, scours are a major cause. Calves cursed with them lose appetite and become so weak they have no resistance to infections like pneumonia which finish them off.

However, even being aware of these facts did not lessen the blow when it came. Poor Plain Jane began to mope, her head hung down tiredly, and she looked miserable and lifeless. Her motions showed the feared diarrhoea. We were going through a period of harsh, bitter weather which did nothing to help the little heifer.

The vet, a tall, saturnine Welshman, injected her without being very hopeful. For him it was something which happened weekly, if not daily. For us it was like having a member of the family ill.

We moved Plain Jane to a separate pen, wrapped her in an old, dry sack and put her under an infra-red lamp we had borrowed. All to no avail. She got weaker and weaker. It was particularly depressing for Shirley who nursed the sick calf, trying to trickle food down the animal's throat. Plain Jane's helplessness and patience upset her and I came in from working in the fields one afternoon and found her crying. She had gone into the pen and found the calf dead.

In the pub they were sympathetic but for them, like the vet, such deaths were a part of farming life.

'You just can't afford to make pets of them,' the understanding Howard said. 'You've got to . . . like . . . stand apart a bit. It's hard but it's the way it has to be.'

Matthew, Old Jonathon's broadbacked brother who had trouble with his legs, said feelingly, 'It's this bloody cold. No use to man or beast . . .'

In an attempt to show that we did not suffer alone, they recalled their own losses or those of friends.

'There was this chap I knew, only a smallholder, nothing much behind him, bought half-a-dozen calves, healthy as you like, fed and bedded them down, come to the pen next morning and they was all dead or going. Every damned one of them.' Matthew paused and looked round. 'Salmonella, that's what caused it. No comeback against anyone, seller nor no one. There's no insurance that can cover you against that sort of thing, Jacky boy.'

'I'd heard that,' Old Jonathon said. 'Nothing you can do but dig a hole.'

We didn't dig a hole. Instead we phoned the Hunt Kennels and a few hours later a young man arrived in a small van. Shirley stayed indoors as he and I carried poor Plain Jane out and loaded her, sacking coat and all, into the vehicle.

'It ain't only you folks,' he said in a rough attempt at consolation. 'This is killing weather. Just you look here . . .' There were four other dead calves in the van.

It was a salutory introduction to the darker side of farming life. Plain Jane's death was the first but not the last for us. There was, as Howard said, nothing to do but put it aside and get on with the job.

13 Planting barley and a gypsy dog

The cows' rapid and enthusiastic consumption of dairy nuts and other expensive, bought-in feedstuffs, not to mention food for the hens and supplementary nourishment for the calves, was recorded in the bills which began arriving from the merchant suppliers. As our local friends put it, 'Money saved on feedstuffs is as good as money earned' and 'Now is the time you should be planting barley.'

Neither Shirley nor I had green fingers. In Commuterland, mine had been decidedly reluctant to get involved in digging, delving and growing things. However, the financial arguments were irrefutable. Even though we did not grow much – just five acres in the field adjoining the house – it meant a great deal to our fragile economy. Besides it was our first arable crop.

Once the decision had been taken things had to be pushed along. Spring planting was still feasible but there was no time to waste. The land had to be ploughed, worked down to a fine tilth and the seed sown. 'It should be in a'fore the end of April if it's to do any good,' we were told.

Thomas, Ellis the cowman's tousle-haired son-in-law, undertook the ploughing: 'Say a pound an acre?' He did not possess a seed drill so quietly-spoken Price, whose farmhouse faced the road entrance to our lane, promised to come along and sow the seed the Farmers' Co-operative would supply; he was a rather shy, hawk-faced young man, usually in overalls and shirt sleeves and apparently unaffected by the weather.

My contribution, apart from the inevitable financing, would be to prepare the ground once it had been ploughed. This necessitated begging, buying or borrowing a set of discs, a roller and spike-harrow.

It meant, yet again, taking advantage of Howard's generosity. A lot of farmers would hesitate to lend equipment, especially to anyone as green as us. Although we were friends it was a lot to ask, but there was no one else. So I screwed up my courage and drove to his place. He was working in the stockyard, struggling with some heavy timber; jacketless, khaki shirt criss-crossed with braces, and rolled-up sleeves.

'Give us a hand,' he shouted almost before I had closed the car door. 'Grab that end and lift.'

There was a lot of strength packed into his short, stocky frame and he was used to manual work. Before the timber was sorted out and stacked I was puffing and sweating. He was amused at something unspoken. 'Bet you wish you'd stayed in that big city,' he said.

I took a deep breath and tried to broach the subject of borrowing but he interrupted and said, 'Come and look at my new calf, tell me if I've been robbed.'

A stocky bull calf, white face and glossy black coat, was lying in a pen built with straw bales in the covered barn. It stood up expectantly when we arrived.

'How much for that one, then?' he demanded.

'Not less than £30,' I ventured hesitantly. My luck was running. It was the right response. 'I got him considerably less than that. £26,' he said smugly. 'Isn't he a beauty?'

It was a very handsome calf.

Having been allowed to praise the calf, I was hurried to the farmhouse still without being given a chance to make my request.

'Look who's here,' Howard said to his plump wife, Dilys. 'Any chance of tea?'

'The kettle's on,' she told him. And to me, 'How's your Shirley and the children? You don't come visiting often enough.'

Howard left his Wellingtons outside and put on carpet slippers. 'She's a devil when it comes to bringing in mud,' he said apologetically.

We sat at the kitchen table to enjoy the tea.

'Well now, he's not come here for nothing, my love,' Howard said. 'Either he's after you or scrounging or stealing. What do you think it is?'

'Not me,' Dilys said. 'At least I don't think it is. Is it Jacky?'

I could feel myself reddening with embarrassment. They both laughed.

'Tell you what,' Howard said. 'Now you'm putting in barley why don't you borrow my tackle to work down that field? There's discs, a roller and spikes.'

Thomas had been there.

'Not very flattering,' Dilys said. 'And I've had my hair permed. He can beg for tea next time he's round here.'

I collected the roller the next morning. It was quite an undertaking. The separate segments which made the eight-foot ring-roller might look solid but they were made of cast-iron and could, very easily, be shattered.

Howard and I coupled it to the tractor and I set off on the two-mile road journey. On the hard surface it made a noise like the coming of the damned. The regulars at the Forge left their lunchtime beers to step outside and see what was passing. I waved but kept going.

Negotiating the rutted, stony lane was the worst bit. The hard, uneven surface threatened the roller every yard of the way and the din was simply fantastic. Animals in the laneside fields took to their heels and got well out of distance before stopping to look and see what they were running from.

A rabbit, goggle-eyed with fright, jumped vertically from its hide in the rough grass of the verge and rushed blindly for the hedge. In its panic it missed the hole it habitually used and ended tangled, helpless and upside down, hanging on the lower branches of a blackthorn bush. Had I dared stop the tractor, it might have gone into the cooking pot but in the circumstances it was in no danger from me. It finally got free, righted itself and rushed through the hedge and away across the field, ears laid back, as if the hounds of hell were after it.

My head was ringing when I reached Egerton; it was a great relief to run the tractor over the turf with the roller, suddenly muted, jingling pleasantly behind. But it took most of lunch to get the noise out of my ears.

After the meal I was off again to collect Howard's discs. Everyone borrowed them and, to a man, they were ungraciously critical. The discs were old, they were rusty, they rattled and shook as if about to disintegrate, but they did the job. Typically I had to collect them from another farmer's fields and was sent on my way with his hopes that it would at least be possible to get them home. 'That Howard,' he declared, shaking his head sadly, 'it's about time he lashed out on a new set. If they was mine I'd be ashamed to lend them to friends.'

They were easier to move than the roller. No fear of shattering and they could be fitted with wheels for towing along the road. Even so it was a relief to park them alongside the roller. By then daylight was fading and it was time to start on the evening chores.

The spike-harrow I left until the following day. It broke down into three pieces and could be, and was, collected in the van.

My final requirement was a chain-harrow and this I acquired in a very dramatic manner. Working round the buildings one morning I noticed an iron link protruding from the rough grass. Just to see what it was I hooked it to the tractor by the tow chain and drove off slowly. A whole square of grass, weeds and turf heaved and billowed and finally disgorged a rusty, holey but useable chain-harrow.

There were holes to be cobbled together with barbed wire, an operation which did my hands no good at all, and the finished article would never have won a prize for elegance in any agricultural show, but it was still a chain-harrow and it would serve my purpose. It was like winning a hefty dividend on the pools.

My leg was pulled when the news got around. Our struggles and adventures generally kept the locals amused.

'You should take a good, long look round about, my lad,' Old Jonathon said. 'With your luck there's sure to be a pot of gold somewhere there.'

Ploughing our five acres took Thomas only a few hours. He drove his big Nuffield tractor like a racing car. I had just sat down for breakfast after getting the milk away when he appeared at the lane gate. Shirley went outside to see if he wanted any breakfast but he merely waved in greeting and turned into the field.

Throughout the morning I could hear his tractor working: the long, droning pull and the change in note as he lifted the four-bladed plough clear, turned and began the pull back. Under the slicing, turning plough the earth opened up, brown and moist and clean. The furrows stretched straight and true from one end of the field to the other.

Behind the plough came the excited birds: yellow-beaked crows, marching regimented starlings, plumed plovers, and tail-dipping wagtails. The upturned earth offered them a banquet of worms and grubs and insects.

It was not the easiest of fields to plough. At some time it had been two separate, smaller fields. The dividing hedges had been ripped out but as the smaller field had been higher than the other, a steep, grassed bank remained to mark where the division had been. This resulted in the tractor leaning over at a disturbing angle at one spot and, since I had to follow with roller and discs, I watched with interest.

Thomas ate lunch with us: a Gargantuan, heaped-plate meal. The conversation was all about cars and engines. He was a star in the local bedstead racing circles. Whether this sport existed elsewhere I never discovered. Nor could I guess what its original inspiration had been. It was very much the younger generation's scene.

Enthusiasts built their own machines, incorporating old-fashioned iron bedsteads into a car chassis and using any suitable engine they could get. Perhaps it all sprang from drag racing, another of Thomas's interests. Whatever its origins, the sport had a dedicated following. There was a league with prize money, trophies and an individual championship.

We went along to watch them once. They raced over a grass track and appeared to reach suicidal speeds. Grotesque machines racing against one another, powered by spitting, roaring engines and flying from one bump to the next. It would have been amusing to discover the reactions of the original occupants of the beds had they seen what was happening.

There was a tremendous interest in cars, motor-cycles and all things mechanical in that part of the country. Instead of leaning over a gate chewing a straw, the average country lad was far more likely to be leaning over an engine with a spanner. Thomas was typical. He had acquired a 3.5 litre Jaguar engine and could hardly wait to shake his rivals with it.

'You see what happens when I get them out on the course,' he gloated and went out to finish the ploughing.

My work began the following morning with the rolling. Driving a tractor on a pleasant day is an enjoyable occupation. From the high seat you get a far view of the landscape. You can see what your neighbours are working at, how their stock is doing, whether their crops are moving.

There was no cab on the tractor. A light breeze moved the air and the sun was warm on my face. No sound of traffic reached me but, occasionally, sunlight would glint on the windscreen of a car moving along a hillside road in the distance. When I stopped the engine to check the roller coupling, a dog barked on a farm half a mile away and the sound carried bell-like and easily but with that content which tells the listener that the noise-maker is not close at hand.

The diesel engine popped away steadily, the roller lurched and thumped over the uneven surface and forced the ploughed earth into a rough conformity. It was exhilarating. There was a sense of achievement.

In the adjoining twelve-acre field a big hare sat up on his hind legs and looked round anxiously when the tractor began working. But when he sensed there was no threat, he relaxed again. Throughout the morning I could make him out, a flattened, khaki stump among the grass. When I came out again after lunch, he had gone.

After rolling, the discs came into use. In quick time the plate-like blades lost their rust as they churned and sliced through the upturned earth. When the machine had been new the discs had been uniform in size, now, worn by years of use, they were all different but they still functioned effectively. Behind the tractor the earth was chopped and reduced until the semblance of a seedbed began to appear.

Later I fastened the roller behind the discs and it bumped along levelling the loosened soil. But there was a long, long way to go before the field was ready. The discs went first lengthways, then across the width of the field and back again. Then it had all to be repeated, chopping the earth finer and finer.

John finished the discing and rolling the next day and went over the field with the spike-harrows until the whole five acres lay smooth and even, waiting for the seed.

Two days after, the taciturn Price appeared, picked up a handful of soil, squeezed it, looked up at the sky and disappeared back up the lane. Half an hour later he reappeared, driving his tractor and towing the seed-drill. His brother, a ginger-haired giant, came along to help.

Basically a seed-drill is a long, compartmented box into which the seed goes. From the box the seed is fed through tubes to the pointed, hollow drills which bite into the prepared soil and deposit it, grain by grain, safely below the surface, in position for germination.

The big brother rode on the bouncing seed-drill, standing feet astride, on a narrow rear platform a few inches above the ground. His weight helped to keep the drills in contact with the soil and his hands were busy keeping the seed free and flowing easily. When Price pulled into the corner of the field to which I had brought the bags of seed, the brother picked up the 140 lb sacks to refill the compartments as easily as if they were packets of potato crisps.

No dilly-dallying about Price. He went off at a sharp pace, taking the drill in an intricate pattern of curves and whirls which might look haphazard but which, in fact, covered the whole surface of the field without time-wasting corners or stops and starts.

As they sped round John followed on our tractor towing the chain-harrow to close the earth over the seed and ensure it was well buried. By mid-afternoon it was finished. All we had to do now was wait.

My next immediate task was to return the borrowed tackle. The spikes and the discs went back without incident. Not so the roller. Right at the top of the lane, within sight of the road, my concentration slipped and the roller dropped into a shallow pothole. There was a sharp crack and one of the segments broke.

One half remained on the roller, the other portion I picked up. Howard was nowhere to be seen so I parked the machine in his field and scuttled off.

'Don't you go worrying any,' Griff at the pub said, resting his elbows on the bar he was polishing and buttoning his waistcoat. His shirt sleeves were held by elastic arm bands. His jacket hung on a hook screwed in the wood panelling behind him.

'No, it's happened before,' he continued, nodding at some memory. 'Damned old roller anyway, almost as bad as his discs. You go and see Holt, the scrap dealer, that old gypsy that gets round the sales. Reckon on paying about half what he asks. Don't let him bite your ear.'

Holt's scrapyard was about six miles from Egerton. I went immediately in the 1800, with the broken bit in the boot. At sales the gypsy bought what others left. The result was a piled-up accumulation of junk and ancient implements and machines, some of which would have been welcomed by a lot of museums. With him they would be broken down for spares or cut up for scrap.

The gypsy was a thin, wizened man in his sixties, perhaps older. His blue serge suit hung on his frame and had obviously been intended for a bigger man. He was playing with a litter of pups, three of them, in a rough kennel made from a heavy crate turned on its side.

'Want to buy one of these?' he asked without straightening up. 'One pound. Be good pet for the kids.' The bitch, a mongrel Labrador, watched me warily. Holt ignored her. 'Here, pick this one up. They'm fat and healthy.'

One eye on the mother, I did as invited. She came forward anxiously but the gypsy told her, 'Lie down, bitch. He's not going to hurt them.' The dog stood still and watched me with her tail waving very slowly from side to side.

The puppy was warm and squirmy like all puppies should be but I didn't want it. 'I'm after a bit for a ring-roller,' I said, 'not a pup.'

He looked from under bushy eyebrows. 'He likes you. You'd be good to him. Take him for seven and six.'

'No thanks.' I put the puppy back and the bitch nosed him reassuringly. 'Griff said you'd find me something.'

'Oh Griff,' he said. 'Come on, then.'

In one corner a ring-roller, about half its segments missing, lay between an iron-wheeled grass mower and a tractor with only one wheel and without most of its engine.

'This do?' he asked, picking up a segment that lay near the roller.

We held the broken bit against it. There was a slight difference in size but it was acceptable.

'How does thirty shillings suit you?' he asked.

'Not very well,' I said. 'How about fifteen shillings?'

He laughed. His teeth were yellow. 'What did Griff tell you?'

'That you'd ask twice what you'd take.'

'Did he so?' he said grinning. 'Let's say a pound then. That's fair and you'm to buy me a pint when we meets somewhere.'

He held out his hand and I shook it. 'That sounds right.'

'Bloody Griff,' he said walking back to the pups. I followed carrying the ring.

When I offered him a pound note he pocketed it and said, 'You'll do well at Egerton. Take one of these pups. Here, I'll give you one for nothing. No, give me a shilling because it's bad luck to give a dog without silver passing.'

I declined the offer and left.

By the time I arrived Howard had discovered the broken ring. 'What's this then?' he demanded. 'Breaking my property, dumping it and running off.'

I produced the segment. 'I've been over to gypsy Holt's yard. Is this OK?'

He was impressed. 'Just come back from him?'

'Straight here.'

'There was no need for you to race off like that,' he said. 'It'll be months before it's needed again. How much did he charge?'

'One pound and a beer and offered me a pup.'

'It's about right. What did he want for the dog?'

'Just a shilling.'

'Aah, that's not for paying, that's for luck. You should have taken it.'

We set about replacing the broken segment and half an hour later the roller was complete again.

It was a lot of hard work and rushing about crammed into a

few days but one morning, some two weeks later, it was all justified. I looked from the bedroom window and, lo and behold, the field was covered with a delicate green sheen. It strengthened daily until, at the end of another week, the whole five acres glistened and shimmered as if it lay under a cloak of finest gossamer. Our barley was through.

 14 Frogs and new grass 

Winter began to break up and loose its hold on the land. Whereas in suburbia we would have looked at the lawn and dreaded to see the day approaching when it would have to be mowed, now we listened dutifully as our farmer friends clasped blunt fingers round pint pots and bemoaned the late coming of the new grass.

It dominated their conversation. They told sombre tales about late snows and sudden frosts that had held back the new growth and so ruined men's planning, and that had, above all, delayed the great day when the cattle – imprisoned in their winter quarters since late November – could be turned out into the fields. These were cautionary tales, told as much to propitiate the shadowy rustic gods as to entertain. The tellers deemed it wise to dampen any hopes that this year spring might come in on cue.

They savoured April snows 'as high as your shoulder' and relived that vintage spring – was it 1961? – when the cold cracked trees and froze streams to the very bed.

From here it needed only a light push by their audience and they were off, jumping decades, to when they were boys and winters were harder and their 'dads' were alive and striding the earth: giants of men who wrought great feats, stronger and wiser than all men who came after them. But whatever the subject, whatever their reminiscences, through it all ran a constant longing for spring and the new grass and release from the back-breaking bondage of the winter routine.

Frogs, not grass, were our immediate problem. The milder weather had brought them out of hibernation. They croaked in the pond in the evenings and their spawn swilled glutinously around in the muddy shallows. They invaded our home.

Egerton, as I have said, was connected to both the mains and local water supplies. We used the latter, which cost us an annual £5 to the local water authority, for all household purposes except drinking and cooking. Following heavy rains it was coming through the taps a rich, muddy brown and there were some free bonuses.

On one occasion a tiny frog plopped out of the tap into Shirley's bathwater and swam two lengths before being rescued and ejected through the window. The kids returned him to the pond and he appeared none the worse for the experience.

Another time Nicholas Paul was enchanted when a fragile shrimplike creature joined him in the bath. Unfortunately this visitor did not survive the shock.

Our nearest neighbour, Willem, and his family shared the local water supply with us but could not understand our concern at the colour of the water. The possibility of swallowing something unthinkable worried them not at all. Their concern was having no water. They were not connected to the mains and nothing was coming through their taps. Past experience pointed to a froggy culprit. On a previous occasion they had been waterless, when a frog for some strange reason crept into their supply pipe and proceeded to puff itself up and very effectively stop anything passing. It was three days before it exhaled and backed out again.

They sought our help and armed with lengths of wire and an assortment of tools, Willem, his elder son (a boy of fourteen), John, myself and the kids trooped along to the collecting tank. It was in the field above our farmhouse. The water was fed into it through a pipe which came from a natural reservoir higher up the slope. This supply tended to be augmented in heavy rains by surface water draining into the tank from the field.

We slid a spade under the protecting stone slab and raised it.

There were indeed two frogs in residence. One was swimming in the water, the second was sitting on a pipe. The swimmer ignored us but the sitter stared back and gulped. When Willem's son lay down and tried to scoop him up he jumped into the water and swam out of reach.

After a quick conference we decided to leave them. Provided they kept out of the pipes, we had no quarrel with them. Willem believed they acted as caretakers, keeping the tank clear of unsavoury visitors like worms and crawlies. (They appeared to have been rather slack in their duties because a big slug lay drowned, bleached and swollen, on the bottom until it was collected in a kid's fishing net and thrown away. Three chickens had followed us and they snatched it up and began to squabble about possession.)

The son managed to push some wire along the pipe taking water to Willem's place and we followed this with several yards of plastic hosepipe – but without feeling anything. Finally Willem stuffed some balls of netting wire into the pipe ends to keep out intruders and we replaced the slab, with an agreement to do something about 'it', whatever 'it' might be, the following day. He had an old well which could meet immediate needs.

We never did. There was no necessity. The same evening, quite naturally, water began running through Willem's pipes again, and he decided it must only have been silt after all.

I tasted the local supply out of curiosity and found it had a certain flavour the mains' supply lacked. As for Shirley, after hearing about the contents of the tank, she was determined to stay faithful to the mains.

All this, frogs, rainwater, everything, it seemed, was a sign of spring and quite suddenly, almost overnight, the fields changed colour. At first it was as discreet as a lady's makeup. The following day it was more pronounced and by the third day it was official. The new grass was through!

The twelve-acre was covered with the finest new growth. Each morning I walked over to look at the miracle and to reassure myself that it hadn't disappeared as mysteriously as it had arrived. In the pub and at auction the conversation now was when the cows ought to be turned out and whether they ought to be risked out overnight.

Opinions differed. They always did. The bold said, 'Take the earliest possible advantage'; the cautious said, 'Husband the growing crop.' But one thing everyone was agreed on was that it was too precious to be fed indiscriminately. It must be strip-grazed.

At some sale or other I had acquired an electric fence kit. It was powered by a small, rechargeable battery. The idea was to confine the cattle to a limited grazing area with an electrified wire which gave a sharp shock to any cow foolish enough to touch it. In this way, as well as protecting the new grass, we should be able to prevent the cows over-indulging in this potent food – with super-laxative qualities – before their digestive systems had adapted to the change of diet.

My dithering about was ended by Old Jonathon when I was collecting diesel from the Forge.

'You got them cows out yet?' he demanded. And when I told him 'No,' he was quite scornful. 'There's a chap at Sollars had them out a week or more.'

'What about the cold nights?'

'Run them in again at night,' he advised. 'You'll have to feed them hay until they leaves it in the racks anyhow. You might as well bring them in.'

The turning-out was an event not to be missed. The whole family escorted the cows to the twelve-acre and the new grass. Our cows were, in the main, an elderly, dignified bunch who'd been through it all before. Nothing we did really surprised them and they tolerated our ineptitude, but this was a special occasion, something to get excited about. Spring comes only once a year and they'd waited all winter for it. So out of the yard they came like schoolgirls breaking up for the holidays, and they made for the open field gate with a rush and a holler.

Before tasting the offering they went pell-mell right round the field, or at least that portion made available by the electric fence. Line astern, shambling along, udders swinging wildly from side to side, bony hips working in all directions. Right up to the fence they went and then turned to run along it. They knew well enough what it was, but even so several pushed their noses against it and got stung for their trouble.

Old Whitey was the odd cow out. None of this spring frolicking for her. Once inside the gate her big head went down, her nose disappeared in the lush feed, and she began cleaning the area around with the efficiency of a vacuum cleaner.

Cows have no top teeth, only a hard pad against which they chew with their bottom teeth. They graze by wrapping their tongues round the grass and pulling it into their mouths. They prefer tallish grass, something to grasp. What we offered them now was several inches high, green and juicy and thickened with clover. It must have tasted like good wine. Not one of them looked up as the family walked away.

By the time I came to collect them in the evening they had cleared the strip allocated to them and foraged along the hedgerows. Everything green had been consumed. They had managed by leaning over or sliding their heads and necks underneath, to steal a good yard on the far side of the electric fence. They lay together, a big raft of black and white cows, chewing away happily, tasting the feast all over again, although from which of their four stomachs the cud was being brought up, I did not know.

On my arrival they climbed lazily to their feet and stretched out luxuriously. Their tails curled over their backs. They plodded through the gate and along to the milking parlour like children coming home tired but contented after an outing. When the milking was over they made their way quietly into the covered yard where the racks were filled with hay. They ate it without enthusiasm.

 15 It's easier in spring 

Hay was another subject about which we had quickly to change our opinions. In the suburbs hay was simply dried grass. Not that it occupied a prominent part in conversations except where some well-heeled daddy sought to impress by moaning about the cost of feeding his daughter's pony.

Things were very different in the country. Hay was feed, hay was money. Any farmer round us could grade and price hay by feeling, smelling or looking at its colour. And the cows could grade it quicker than the cleverest farmer. They simply walked along the racks and sniffed. Their noses told them all they wanted to know, and they ate the choicest bits first in the certain knowledge that, if they did not, the next cow along would.

The arrival of the grass transformed our workday lives and routines. The cows still needed hay, but as the days passed and they made the changeover, more and more was left in the racks. Even so, it was well into May before they were completely on grass.

There were lots of benefits. Dairy cows on fresh grass don't need expensive dairy nuts and we simply gave them a few handfuls to secure their cooperation when milking. When they began lying out in the fields at night they came into the milking parlour with clean udders. A great improvement. Not so much work washing them in the mornings. There was a saving on bedding straw. The workload, at least where the cows were concerned, was considerably lightened. There was almost, but not quite, time to stand and stare.

One distinct disadvantage of new grass was the traumatic effect it had on the cows' bowels, especially in the early days. It acted like some super laxative. The milking parlour was awash with the sloppy green stuff. A hiccuping cow, and they often did hiccup, was a threat at six yards' range. After each milking the walls as well as the floor had to be hosed down.

But it was all worth while. Our world was green. Hedges were renewed, trees were leafing, bank and verge glittered with dandelion gold, the bracken in the lane began to unfurl, and the tall spikes of foxgloves thrust up through the dead, tangled débris of winter.

Pheasants began mating. These gorgeous birds were bred on the nearby estate and there were a lot of them. Driving the tractor I would surprise strutting cocks lording it over retinues of more drably feathered hens. One cock bird was particularly beautiful: his feathers were almost black. Several times I came on him in the lane and scattered his harem of adoring hens.

Tiny rabbits began to pop up around the farm. Myxomatosis was still with us. On one occasion, visiting a neighbour, I nearly ran over a rabbit creeping blindly about the road. His head was swollen with the disease, his eyes were puffed, closed and sightless. I stopped the car and put him out of his misery.

But there were other rabbits. Quick, lively, jumpitty rabbits. One fat specimen used to lie in a shallow hide in the lane hedgerow. Most mornings he was there when I took up the milk in the transport box we bought at a sale and fixed on to the tractor. When he heard me coming he crouched and flattened himself and stared at me, apprehensive but not scared enough to bolt. I plotted how to transfer him from the hedgerow to the kitchen, but one day he was not there and I never saw him again. There was an old red fox that worked the area, perhaps he could have explained what had happened.

The dairy herds began calving to catch the spring grass and that meant more calves on offer in the auction rings. The calf-dealer came twice in a few days. First he brought a quartet of black heifer calves at £12.50 each. They were rather thin but healthy enough and they took hungrily to the bottle. Our labours in feeding calves had been lightened by neighbour Willem. He showed us how to convert discarded rubber milking teats into bottle teats for the calves; it saved us hours of work with buckets and probably also saved a few calves.

The calf-dealer's second visit brought Ferdinand into our lives. He was the odd one out in another quartet. There were three sturdy Hereford Friesian heifers with that rusty red colouring that turns almost to black with maturity, and Ferdinand, a handsome, red bull-calf. The asking price was £20 each. The locals thought that a bit high but we had no hesitation. We were anxious to build up our stock and these were heavier and stronger than the other four. In calves, as in most things, quality has to be paid for.

It did not take us long to discover the striking difference in temperament between heifer calves and bull calves. The former are almost always quicker to learn and far more independent. Bull calves are slower and tend to be slightly stupid but they are also much more affectionate. Ferdinand was like an outsized kitten and was spoiled like one. He loved it. Given the chance, he would wind himself round your legs like a hungry house cat. When we loosed the calves into the stockyard on sunny days, he galloped up confidently to be petted and then followed us around, reminding us with an occasional nudge from his nose that he was there.

The amazing thing was that he never changed much. Throughout the time we had him he would hurry up to greet us as friends. Even when he had grown into a hulking bullock his need for affection remained. The kids adored him. When he was big enough they rode on his back. One young Londoner visiting us spent an idyllic week playing with his bullock friend and could hardly be induced indoors for meals.

This capacity for friendship among calves and grown cattle was intriguing. Calves reared together tended to stay together for the rest of their lives. Perhaps it was the herd instinct. Whatever the explanation, their friendships endured. Even when they were old enough to be turned out into the fields they still grazed together as a group. If they were put in with twenty or thirty other cattle they stayed together. What might appear to be a big herd was usually a collection of small groupings.

Ferdinand's size and high spirits singled him out in the fields. When life was good he gambolled like a lamb, kicking his heels in the air and racing about with his chums like a street urchin. When things were not so good, he needed reassurance and came up to be petted. Had it been practical to keep him as a pet, we would have done so. But we were false friends. Farming and sentiment do not mix and Ferdinand was a beef animal.

 16 Our first calf is born 

The first calf to be born on our farm arrived at the end of April. The cow was the young, wild-eyed creature I had bought at the dispersal sale and it was probably her third calf.

It began for us when I went to bring the cows from the yard for the early milking. At this time the cows were going out to grass during the day but were still sleeping in the yard at night. They queued at the door and the minute it was open they began trooping through. Fortunately the young cow was at the end of the queue and I spotted the calf before she could get through the door.

A first reaction was panic. We knew, of course, that the calf was due soon but its actual arrival was still something of a shock. I had never seen a calf born. The nearest I came to something like it in Commuterland was the cat having kittens – three of them – under John's bed in the small room. Now here I was with a calf on the way and no one to run to for advice.

Although part of the calf was already visible, the young cow seemed unbothered. We had been bringing her into the milking stalls with the others, partly to get her accustomed to our routine, partly to 'steam her up', which meant feeding concentrates to stimulate her milk flow and ensure a good lactation. Now she was eager, calf or no calf, to follow the others and reach those dairy nuts.

I turned her back and closed the door. Next need was to try and manoeuvre her – she was very suspicious – to try and identify exactly what part of the calf was showing. It proved to be the front feet with the nose coming through above them. That, to my great relief, was according to the book.

Some of the panic subsided. Old Jonathon never wearied of telling me, 'Leave 'em be if they'm happy. Cows 'as been having calves a long time. They knows a lot more about it than you.' Now I followed his advice and left her in the yard while I got on with the milking and tried not to worry about what might or might not be happening.

It was impossible to concentrate on the work in hand. Every few minutes I had to break away and run for a quick look at her. Nothing appeared to be happening. Each time I appeared in the doorway the young cow made a dash towards me, hoping to be allowed to rejoin the others.

The big event was delayed until I had helped load the milk into Jock's lorry and seen it taken away. I raced back down the lane and parked the tractor. By the time I opened the yard door it was all over.

The calf had arrived. Birth-wet and limp, it lay in a patch of sunlight. This time the young cow made no move towards the door. She was busy licking her calf dry. The big tongue roughed and cleaned the calf's coat and the new mother made low, soothing sounds to reassure the little animal.

When I approached she looked up apprehensively and lowered her hornless head but it was a formal warning. She knew me, knew my scent, and I was acceptable and represented no threat to her calf. Even so her eyes were anxious as I knelt to examine it.

There was really nothing for me to do. Nose and mouth were free of mucus. The calf was breathing easily and already lifting its head and beginning to stretch its limbs. It always amazed me how quickly new-born calves grew strong enough to stand and move about.

It was a handsome heifer. Beautiful, glossy black coat with white underparts and the mandatory white face. Its limbs were straight and strong. In short, she was perfect. In dairy heifers sometimes there are supernumerary teats which have to be removed – scissors are recommended – if the calf is to be reared for dairying. There was nothing wrong here, and anyway this youngster would be raised for beef.

I left the cow to get on with it and went inside for breakfast. My first instinct was to blurt out the news but the chance to playact was too great.

The kids were home. It was a Saturday. Breakfast was sizzling on the stove: Shirley always started it when she heard the tractor returning down the lane. I washed hands and sat at the table.

'What kept you?' she asked. 'Anything wrong?'

'No, nothing wrong.' I got busy with knife and fork. 'Just a new calf.'

The kids joined the act. 'A new calf? Where?'

I tried to cool them down a little. 'Don't rush. It's just been born in the big yard.'

It was a waste of time continuing. I was talking to an empty room.

The hay bays which formed one side of the big yard gave a good viewing platform to see cow and calf. The cleaning-up process was still going on, and the cow was trying to nudge the calf to its feet. In the sunlight the calf's damp black coat glistened. When the kids exclaimed, the cow looked up at the gallery, snorted, and returned to the task. The kids gave me their promise to do nothing to disturb her, and Shirley and I went indoors to a now cold breakfast.

For us the calf was a welcome bonus in the struggle towards solvency. For the two youngest she was a miracle. They sat entranced, discussing each movement in low whispers.

It was quite a struggle before the calf managed to stand up. Several times she rolled right over and lay temporarily exhausted before the anxious cow coaxed her into having another try. But at last, much to the kids' relief, she succeeded and balanced precariously on trembling legs.

The worst was over. Five minutes later she was nuzzling and butting at her mother's udder until she succeeded in drawing the teat into her mouth and began to suck. The cow stood contentedly. When the calf had taken her fill, she lay down and slept, curled neatly into a ball, while the cow stood close at hand chewing the inevitable cud.

It would have been nice to have joined the young ones but farming chores involve hungry animals and cannot be shelved. So first I had to help John and Shirley feed the bigger calves and wash the tack. Afterwards I went into the yard to take a longer look at the newcomer. There was no sign of the kids but when I bent to look at the calf, Vicky's voice floated out from her hiding place among the hay bales.

'We are secret service agents and we are spying on that little cow. You have been warned . . .'

Before lunch the cow had cleansed herself, which meant clearing the afterbirth. I picked it up on a muck fork and buried it in the dung heap to rot.

Sometimes, as we discovered, the cleansing does not come away naturally. This means an unpleasant job for the vet because by the time it is ready to remove it is ripe enough to smell and look foul. When our vet's young assistant did the removal he used to strip to the waist and put on a long rubber apron. Afterwards, no matter what the weather, he washed in the warm water we supplied and rinsed off in the ice-cold local supply in the stockyard.

We held a family conference to choose a suitable name for our new arrival. Prima was agreed because it was the first calf born on the farm. The cow had been previously called 'Wild Eyes' but henceforth was known as 'Mother'.

We left this first calf with the cow for a week but experience showed it was best to allow the new calf a good feed of colostrum – the first milk produced by the cow after calving – and then to separate them. Afterwards the cow was milked into a separate churn and the milk bottle-fed to the calf.

Colostrum is a precious commodity. It is vital for a calf's health and chances of survival: it contains antibodies which protect the newborn against harmful bacteria; it also contains essential vitamins and has a laxative effect which cleans out the calf's intestines. It is quite distinctive, sometimes a definite reddish pink colour. We stored surplus colostrum in plastic containers and deep froze it to feed to new, 'bought-in' calves. The system worked well and was no loss to our milk sales because we could not send the mother's milk to the dairy until the colostrum had cleared – usually about half-a-dozen milkings after she had given birth.

Prima thrived and was allowed to go into the fields with her mother. No matter how sharp the weather, the cow always seemed able to find a protected, wind-free spot to settle the calf before going off to graze. But there were complications.

These came in the form of another cow, Ermintrude, who

must have set high store by motherhood. Normally she was well down in the cows' pecking order and extremely reluctant to intrude on any other cows' rights, imaginary or otherwise. If we promoted her in the milking rota so that she came into the parlour out of turn, she had to be driven in and was very unhappy. Now, here she was, completely out of character, showing an aggressive streak with the clear objective of gaining possession of the calf. Quite deliberately she would push between Mother and calf and try to lead the baby away. The worried cow was continually chasing after them to re-establish her maternal rights. For the kids the would-be baby-thief became 'Aunt Ermintrude'.

The struggle came to an end when we took the calf away from the cow and put her on bottle-feeding with the others. She quickly settled into her new way of life but Mother and Aunt Ermintrude bellowed themselves hoarse calling for her. Even after Mother had accepted the inevitable and resigned herself to the loss, Aunt Ermintrude persisted in calling and for a couple more weeks we had to watch closely or she would sneak away and search for the little one.

We later discovered that if a calf was taken away from its mother soon after birth, a cow tended to settle down quickly and forget. If she was allowed to keep her baby any length of time it became a very different affair. She would bellow and call for hours and hunt for the calf, even trying to sneak in from the fields. Several times we chased away cows trying to butt down a door to reach the calf inside. Heartless though it might seem, early separation was best. In any case, some of the oldest dairy cows proved rather indifferent mothers. Perhaps their maternal instincts had been blunted by always having calves taken away.

17 Calves, bulls and bowler hats

Unfortunately not all calvings were as easy and uncomplicated as our first one. Later in that same year I bought a big but very nervous cow which we called Jemima. She was in calf and one day in late summer she found a quiet corner of the top field to produce. Now the difficulties arose. She could not deliver the calf; even though the head and forefeet appeared they were drawn back again.

Eventually I went to seek advice and help. Our shy neighbour, Price of the beak nose and few words, was working with a tractor near his farmhouse at the top of the lane. He listened, his head held on one side like a bird, simply nodded, climbed into my car and said, 'Let's go take a look at her, then.'

The cow had been trying to bring the calf for some hours. Before leaving I had put her in the collecting yard and now we ran her into the milking stalls and fastened the restricting chain behind her. Price stood on a straw bale and inserted a carefully scrubbed and soaped hand.

'The head's there right enough,' he said cryptically. 'But her's got to come from there quick or we'll lose this calf.'

Under his instructions we built a floor of straw bales to protect the calf from injuring itself on the concrete floor at the moment of birth. Next he took two short lengths of thin rope and tied bits of wood to them to give us something to grip. Reaching inside he pulled out the forefeet and tied one rope to each foreleg. A lot of of the natural birth lubricant had dried round the vagina, so we substituted a wet soapy solution.

The cow was still spasmodically trying to eject the calf and Price timed our efforts to coincide with her rhythm. At his signal I began to pull on the ropes while he guided the calf's head and feet into the proper position. 'Once her starts to come proper, keep her coming, don't you stop pulling until I says stop,' he warned.

Pulling calves is excellent training for soft stomach muscles. Mine were trembling before the cow groaned loudly and the calf's feet and head re-appeared at last. Price took one of the ropes from me and we both lay back and pulled like ancient mariners hauling in the anchor. Even so I began to think we must fail until, quite suddenly, the cow went down kneeling on her front legs, bellowed her protests and suffering, and gave one last mighty convulsion of her muscles. The head came clear, and then the shoulders and, in seconds, two-thirds of the calf was out. I would have slackened, but Price shouted, 'Keep pulling. Get his hips clear. Keep pulling.'

There was a further short resistance but then the calf, an enormous baby, slithered to the straw bales. To me he looked lifeless, exhausted with the effort of being born, but Price thought otherwise. He went down on his knees working, clearing the nose and mouth and calling for cold water. I brought a jug from the dairy and he sprinkled some on the calf's head and neck.

The shock did the trick. The calf trembled and then drew in a harsh, painful breath. His chest began to move, shallowly at first but then more strongly.

'By God,' Price said, looking up at me, 'he's one of the biggest calves I ever saw. He's a good 'un all right.'

A normal Friesian–Hereford cross calf would weigh, I suppose, something around 80 lb at birth. This one must have been 15 to 20 lb more than that.

It seemed advisable to give the calf time to recover from the birth struggle before moving it to a pen. We laid straw in one corner of the milking parlour and carried the limp creature over to this bed.

The cow was on her feet again and trying to turn round in the stall.

'What about her?' I asked.

'Now we must get her out of there,' Price said cheerfully. 'Either she'll take to him or her'll go mad. I've known them go proper daft after a difficult birthing.'

Not this time, however. The cow took willingly and quietly to the calf that had caused her so much pain and trouble. She sniffed him carefully, decided he was indeed her own, and began licking him into life.

Price and I washed up. It had shattered me but he took everything in his stride. In his quiet voice he recalled the time when they had to use a block and tackle and a Land-Rover to 'get' a calf. My thanks embarrassed him. Not even a cup of tea would he take. Instead I ran him back up the lane to where his tractor stood. Before I could turn the car he was working again. Not me. I drove back with shaky hands and went inside to sit down with a coffee, report the new arrival and recover.

The calf got over his bad start and grew like a mushroom. He was the product of one of the top Hereford bulls and he showed it. Probably he was the best animal we ever produced on the farm. The kids called him Charley-Baby.

One cow bought for £92 at a hill farm sale proved a marvellous investment. Less than three months later she produced a whopping big black bull-calf which we sold because we needed money even more than usual. Five days old, he went for £75. The trade collapsed shortly afterwards, and a year later he would not have fetched as much even though by then he was a strapping, hungry bullock.

One fact of farming life quickly appreciated was that dairy cows produce milk to feed calves, not fill bottles. Having given birth they begin a 305-day lactation during which they produce varying quantities of milk, depending on the individual cow, before going dry until they calve again.

Cow management is essential if you are to survive as a farmer. In theory we should have regarded our little herd as a collection of milk-producing units. But with my family there was no way that could be done. The cows fell into a category somewhere between pets and animal friends. Machines never! One cow we acquired would sometimes stop grazing and come up to be petted. It made it difficult for us to see her as a milk-producing unit and, somehow, I think our attitude made it difficult for our neighbours to see us as 'proper' farmers.

Ideally a cow should produce a calf every twelve months. The period of gestation is 280 days and once the calf is born the cow comes on heat every three weeks until she is in calf again. The recommended practice is to put them to the bull about eleven or twelve weeks after they give birth.

Our economy did not permit us to keep a bull, so we used the 'bowler-hatted bull', which was the local tag for the artificial insemination service. The main problem with this is that although a cow comes on 'bulling' for three days every three weeks, she is fertile for only part of this time. If you miss this period, there is nothing to be done but wait until the next time. Thus it is important to be able to judge just when artificial insemination will be effective.

However, when a cow is 'bulling' there are usually plenty of signs. Her herd mates get excited and often try to mount her, playing the part of the bull. Some cows will actually call the bull, stretching out their necks and bellowing for the master of the herd to fulfil their needs. There was something of the voyeur about it all, but then sex plays a big part in livestock farming. All the family got used to looking for the signs. Even the kids would inform me, matter-of-factly, 'Such and such is bulling, you'd better call the AI man.'

We used the Milk Marketing Board AI service; by this means even small farmers such as ourselves were able to use top-class bulls we could never, otherwise, have afforded. It was a tremendous boon.

They gave us a telephone number to call and we had to specify what breed of bull was required, whether it was to be the 'bull of the day' or a 'Star bull' (which meant paying a few pence extra). In our first few months we had a fair percentage of failures on the first service which meant we had misjudged the brief fertility period; but the service allowed for a number of free repeats and, with experience, our judgement improved. The basic cost of the service was £2.40, with the 'Star' bull entailing an extra 30 pence. Our reasoning was that the extra charge gave one the choice of the country's best bulls and was money well spent.

The AI men kept the precious bull semen in straws which were carried in a portable freezer container in the boots of their cars. The straw fitted into a hollow rod with a plunger which forced the semen out when the operator wished. The skill lay in placing the semen correctly to achieve fertilization.

Afterwards the operators issued a certificate naming the bull used and particulars of the cow and dates. The cost was deducted from the monthly milk cheque. It really was simplicity itself.

A few important breeds, mainly French Charolais and Simmental, were used by our neighbours to see what sort of calves resulted, but in the main the tried and trusted Hereford remained the favourite bull to cross with Friesian dairy cows. It was intriguing to learn that because of the scientific storage of bull semen, a prize animal could sire hundreds of calves long after he had died or gone to the abbatoir and been turned into beef.

Bulls have a prime role in rural mythology. They were a favoured topic of conversation among our friends. They are animals apart, symbols of fertility and increase. There is something very majestic about the great, heavy, groomed and coiffured bulls paraded for the judges at the agricultural shows. Even the workaday bulls that never leave their home farms have something of this quality about them.

But handsome as they are, the general feeling among the men who work with them is that they are also pretty dim. It seems to be a general feeling that vicious bulls are few and far between, although almost all the men we knew could tell a story about someone who had been killed or maimed by a bull.

'They'll do something daft in a split second,' a man named Hughes said one darts night at the Forge. 'There's no accounting for them. This Friesian we had we'd reared from a calf, a beautiful creature. Something must have happened. One morning when I was just a bit of a lad I went out into the yard and there he was just standing. The pen door was open behind him. so I walked over, slapped him on the arse and put him back. In he goes, quiet as you like, but when I go to shut the door, there's poor old Arthur, our cowman, dead as a doornail. Ripped wide open. Yet that bull was gentle as a lamb. None of us could work out what might have happened.'

'What did you do about him?' I asked.

Hughes shrugged. 'What could any of us do? He had to be

put down. We just loaded him up and took him along to the abbatoir the next day, as soon as they could take him in. But it never answered anything.'

It is this element of uncertainty, coupled with their tremendous strength, which makes bulls dangerous unless handled correctly. One toss of their head, and with that powerful neck anything could happen.

Another tale told the same night was of the bull that lifted five heavy gates off the hinges, hooking them up with his horns, to reach a cow calling for his services in a nearby herd. 'No good any of us trying to turn him once he'd got the smell of her,' the teller said. 'He'd have walked through a brick wall to get there.'

Personally I found Jonas's story the most amusing. Standing in the Forge in his crumpled suit, stooping to listen to his shorter friends, he looked far from being a successful farmer/trader. But the long, lugubrious face and deceptively heavy-lidded eyes masked a shrewd mind. It was always nice to see him walk in because he had an inexhaustible fund of anecdotes. This one concerned some cross-bred heifers he wanted to breed from and use in a single suckling herd. He tried to hire his rich neighbour's prize Hereford bull, only to be rebuffed.

'Silly old fool,' he said. 'They wasn't good enough for his precious bull. We'd a few happy words about it, so I just went away and bided my time.'

At a very opportune moment Jonas put his heifers in a field on one side of the river which divided the two farms. It so happened that the bull was with a herd on the far side, and when the heifers bellowed their needs, he, being a well-bred gentleman, swam the river and offered his services.

'You should have seen that old boy's face when I went round there and complained,' Jonas said. 'He almost choked when I told him to get his bloody bull off my land and stop him molesting my heifers. That bull had been there for three days then. Charlie Haines, his foreman, said he nearly had a fit when he got the letter from the lawyer saying his bull had got my cows in calf and demanding compensation. But he had to pay or we'd have taken him to court. Mind you, I wasn't after the money, it was the principle of the thing.'

No one tried to follow that story.

 18 The two porky sisters 

Despite the warnings of economic disaster, we acquired two sows. The farm had been planned as a small dairy pig unit and we were quite well equipped with buildings for weaner or porker production.

Local opinion, when consulted, was mixed.

Old Jonathon's man, Billy, who had a thing against pigs, was gloomy. 'Eat you out of house and home, they will,' he warned. 'Nothing in them. Eat you out of every shilling you've ever owned.'

'Nice to have a bit of pork about the place, though,' Griff said.

That Billy could not deny. 'Eat them before they eats you,' he said shaking his head.

Old Jonathon spoke up suddenly, 'I like a nice bit of belly pork. Good, crisp crackling.'

The truth was that we had decided to 'try' pigs as much for experience as anything else. Neither Shirley nor I knew anything about them, so our knowledge, one way or another, was certain to be enriched. We consulted our books but we were rapidly learning that the written word needed to be backed up by practice and, after much discussion, we agreed to have a go.

Our two pigs, bought from a local farmer, were sisters from the same litter. They had been brought up together but in character and appearance they were very different. We named them jokingly after two friends – Dorfie and Dorrie – and were horrified when these ladies visited us and the kids blurted out the news. Fortunately their sense of humour overcame their initial indignation.

These porky sisters were young Large White/Landrace cross-breds. Each had had one litter before we got them. Dorfie was the biggest. Definitely the washerwoman type. She liked nothing better than a good grunt with her owners and was always hanging around people to see if anything edible was likely to turn up. And good-natured or not, she never hesitated to bully her sister off any food that was going.

Really, the second sow, Dorrie, had much higher standards. For instance, she did not make the same disgusting noises as Dorfie at mealtimes. Nor was she in the habit of rattling the door to the feed-store in the hope of shaking it open. Nor did she try and brazen things out when caught in some forbidden act. Not her. She laid her ears back against her head and ran for it as hard as her legs would take her.

Several of our misconceptions about pigs rapidly vanished. Far from being dirty, they are more fastidious than most animals. They are comfort-loving creatures and much prefer a dry, well-strawed pen to the mucky, messy pigsty tradition bestows on them. 'Pigs is as clean as their owners,' Old Jonathon said. 'They likes muck well enough but not their own.'

Gluttony was their vice. They guzzled whatever was offered and looked for more. Our usual route to their pen was via the collecting yard. They knew the feeding times and were on the alert for the sound of buckets; they would stand with their fore-legs resting on the front wall of the pen, twisting their heads to see us. When we did come into sight, they almost exploded with excitement, screeching and grunting in anticipation. By contrast, they would drink clean, fresh water as delicately and appreciatively as a maiden aunt sipping her favourite brand of China tea.

They were certainly sociable. Dorfie in particular would greet us with obvious pleasure. She had a special, rather explosive, short grunt which was a greeting sound. There was no mistaking it. Both liked to have their backs scratched and would wriggle about to make sure we reached the right spots.

Pigs are as intelligent as dogs. Their powers of smell are staggering. No wonder the French use them to find truffles. Years ago I read about an eccentric old squire who reared a piglet with a litter of Pointer puppies so that it grew up believing itself to be a dog. When it came to finding game, the pig left the dogs standing. The owner used to back it against his friends' fancied dogs and pick up some easy money.

It was easy to believe the story when we saw pigs in action. They were always on the lookout for anything edible: animal or vegetable, it was all grist to the mill. They homed in on rubbish and carrion like something out of a science fiction novel.

Before they had been with us a couple of weeks they knew exactly where to go to supplement their rations. Young nettles were a favoured titbit. They first ate the tender, growing shoots and then nosed out the roots and polished them off as well. Where there had been beds of nettles, they left nothing but an expanse of well-worked soil.

Dorfie's private hunting place was the farm fire where we burned rubbish. It was in the spare lot by the pond and provided all kinds of delicacies for her but sometimes added a few shocks as well. On one occasion a great outcry of squealing and grunting had me chasing out of the dairy. The site of the fire was almost obscured by smoke and flying ash. The previous day a chicken had died and we had thrown the carcase, feathers and all, on the fire. The feathers burned but the rest, well, most of it, remained cooked to perfection for our porky gourmet. But she had run into a snag. The embers were still smouldering and the bird was so well done it came to pieces when she grabbed it. So she kept dashing into the scattered fire to snatch a mouthful, screeching when her nose and toes got scorched, shaking her head from side to side to ease the sting. However, it must have been cooked to her taste because she persisted until everything, bones, feather stubs, the lot, had gone. Then off she went to cool her nose in the moist mud at the pond's edge. Probably it was the high spot of her scavenging career.

Shirley was continually at war with the pair of them. As the weather improved, the hens began straying and laying in the hedges, in clumps of grass, anywhere their fancy took them. And pigs, we discovered, love eggs. Their noses enabled them to track erring hens to hidden clutches and they gobbled up the eggs, shells and all, with great relish. The sight of them near the henhouse or in 'chicken territory' had Shirley dashing outside, waving a stick and threatening them with extinction at the very least. Off they would rush, like children caught in the orchard, much too fast for her to get near.

They always returned, particularly after Dorfie discovered that the henhouse door could be opened by rattling it if someone had forgotten to push home the bolt. To get into the henhouse itself was as good as a birthday. There were marvellous pickings: eggs and food pellets the hens had missed. In their enthusiasm they created chaos with the nesting boxes. Noise was usually their undoing.

Another *verboten* area was the garden. If they could get in they uprooted flower bulbs and ate them. Whenever the gate was left open they would try and sneak through but they had to pass the kitchen window and a tap on the glass was sufficient to send them scurrying.

Of course, the reason for acquiring the pair was not to study pigs' social behaviour or compile dossiers of their crimes. Nor was it our intention to add a couple of outsize pets to the menagerie. The idea was to produce more pigs and that meant we had to obtain the services of a boar.

After weaning a litter, a sow usually comes on heat in between three and eight days. If they are 'missed' they come on again at three-week intervals. One of our worries was whether we would recognize the heat symptoms.

'No need to bother about that,' the selling farmer assured us. 'They'll let you know well enough.'

Strangely, although it sounded unlikely, that was exactly what happened. Dorfie Pig simply marched up to me in the stockyard one morning and proceeded to say in a series of very agitated grunts that something was very, very wrong and needed to be rectified.

So, out with Old Lil and in with the pair of them. No problems. They jumped in willingly, following Andy, a Scots friend staying with us, who was carrying a bucket containing a few pig nuts. They ignored him when he scrambled out again and closed the doors. They were too busy quarrelling over the bucket and its contents.

Howard had a Large White boar. It was a huge animal. When we arrived with our two ready and very willing sows, the boar was with some other pigs in a roadside field. Howard was feeding them, tipping ground barley from a bucket on to the grass. Now a strange thing happens when a sow in heat is brought to the boar. They simply 'freeze' and wait to be served. It happened here. We might have been advised by Dorfie but it proved that Dorrie was the one most in need. She waited apprehensively as the boar approached and sniffed but his lordship had priorities. The sow would wait. The barley would soon be eaten. He went back to make sure of his share.

Andy was awestruck. 'I hope I never get that hungry,' he said feelingly.

We left the sows with the boar. When we collected them on the third day, the boar's fatal charm had evaporated as far as they were concerned. The sight of the van and us with a bucket brought them running, and they jumped in readily for the journey home.

Howard charged me £1 a sow. 'Just for the cost of their feed,' he explained apologetically. 'It's half what I'd charge anyone else for the boar service and there'd be extra for the feed.'

Now they had been mated we had three months, three weeks and three days to wait until the piglets arrived. In the meantime, nature's urges satisfied, the sisters settled back into their old routine of scrounging, thieving and stealing.

Someone – Thomas, the cowman's son-in-law, I think – pointed out that the bumpy, little triangular Camp Field at the top end of the farm had good hedges and wire fencing and was supposed to be pig-proof. There was a little stream to provide water and a sound, dry shed for the sows to lie in at night. Why not put the pair up there? It sounded good advice, and so one bright May morning, plus the inevitable bucket and pig nuts, I set off.

The sows were delighted. 'Walkies with Daddy' was as good as a Sunday outing. They gambolled alongside or at my heels full of excitement, with Dorrie occasionally trying to snatch the bucket out of my hand. Once in the Camp Field I deposited the nuts in a trough previously carried there and left, closing the gate behind me, as the pair guzzled the offering. A simple enough operation. The pigs were in the field.

They were in, but not for long. Next morning I found them lying contentedly in the cow yard with a circle of incredulous, disapproving cows standing round them. They were delighted to see me, grunted their welcomes and sniffed around in the hope that I had brought them something to eat.

Later that day, after repairing a hole they had torn in a stretch of wire netting near the gate, I tried again. The littles came along this time to see the fun. 'Pignicking' was how Vicky described the excursion.

This time the pair knew the game. They bolted the goodies in the trough and rushed after us. At the fence they simply put their noses under the wire, lifted, and slid underneath as to the manner born – as, indeed, they were. In the wild state, so John who read these things informed me, pigs are creatures of bush, scrub and forest. They have the perfect bullet shape for penetrating undergrowth. If this be so, what chance had our flimsy wire fences? We conceded defeat and went back to the old arrangement.

We learned a lot from the two of them. It was surprising to us to find that although they needed cereal foods, they would also eat grass in the fields. One difficulty was that they rooted up the sward and left divots of turf all over the place. This could not be allowed to continue. Grass was precious.

'No wonder,' Howard said. 'They've lost their nose rings. You'll have to ring them again. Have you ever done it?'

I assured him there was not much demand for ringing pigs in London.

'Then I'll give you a hand,' he said.

Ringing, like many operations in farming, is not for the squeamish. The copper rings come open and sharpened at both ends. They fit into a groove in special pliers. 'All you have to do is close the ring on the pig's nose.' Howard said. 'It's just like women having their ears pierced. Nothing to it.'

I was prepared to take his word with certain reservations. It was not yet clear to me how we were going to hold 350 lb of pig, perhaps more, still enough to be able to do this small thing.

The answer proved to be a simple slip knot. Howard made one out of thin rope, dropped it deftly over the unsuspecting Dorfie's snout, and pulled it tight. 'You've got to get it behind their tusks or it'll come off,' he explained ignoring the distraught screeches from the pig. 'That's the secret.'

Poor Dorfie! That such an outrage could happen to her! The odd thing was that she did as everyone had said she would, pulled back on the rope which we fastened to a post, and squealed fit to bust. It was for all the world as if the rope had been transmuted into a stiff metal rod. Not once did she attempt to come forward and reach either Howard or me.

Even so I kept a wary eye for the first sign of aggression. Not so Howard. He ignored the noise and began, calmly and efficiently, putting in the rings, closing them through the tender nose, positioning them to prevent the sow rooting. When it was finished, he released the knot and freed her and the sow dashed thankfully into her sleeping quarters to nurse the injury.

In the next pen Dorrie was frantic at her sister's squeals. Had there been no dividing wall, she would probably have attacked us. So, when we entered her pen I was prepared for trouble and armed myself with a muck fork. It was not needed. Her courage evaporated when she found herself threatened. She tried to dodge the noose, holding her head close to the floor, but there was no escape and we had a repeat performance, except that Dorfie was less responsive to her sister's protests than Dorrie had been. When we left, she put her head round the corner of the inner pen and looked mournfully at the friends who had so ill-used her. But she made no attempt to come out.

It was a pity, but there was no alternative if the pair were to be allowed out. For a few days they nursed their tender noses but before too long they were snuffling about as usual and we were getting the same good-natured greetings when we met. It was nice to be back on the old footing.

19 A crippled hare and a flamingo

Now that the grass was coming through the meadows had to be chain-harrowed to rip away the dead, matted stuff and allow sun and air to reach the new growth underneath. It was easy, pleasant work. The tractor diesel popped steadily and my old, raggedy, much-laughed-at chain-harrow rattled and seethed along behind, picking up and bringing with it an assortment of bits and pieces, stones and rubbish.

It was astonishing how in the barest of fields a few minutes' activity with the chain could produce quite hefty rocks, chunks of timber and even red housebricks. They must have been embedded in the earth and not visible until ripped up by the chain and brought into view. No point in leaving them in the field, so every now and again, whenever it seemed worth while, I stopped the tractor, climbed down and carried an armful of whatever was offered to the side and tucked it away in the bottom of the hedge.

This day I was working in the fifteen-acre; a long, triangular-shaped field which sloped steeply down to the stream marking the farm's southern boundaries. It was interesting ground with a variety of undulations and a pit which might have been a dew pond. On some sections the tractor heeled over markedly which added a little spice to life. My predecessor at the farm had on one occasion ended up in the stream which ran along one side of the field, tractor and all, when his brakes failed.

An isolated, stumpy oak stood in the centre of the field. It was the sole remanant of some long-gone dividing hedge. Beneath it the grass was coarse and rough and when the work brought the tractor near to the tree something stirred in this uncultivated patch. The movement was repeated on the return run, so I stopped to investigate.

When I walked up a big hare lifted from where she had been lying flattened against the turf and watched me. I was within two or three yards when she made an attempt to run, but the days when her legs would have taken her out of danger were over. She covered maybe six yards and stumbled and fell. She was lame.

When I approached again, she made a second attempt and floundered about pathetically, though still covering ground quickly enough to convince me I would never catch her. She lay awkwardly out of my reach, her big eyes watching for any movement on my part, ready, notwithstanding the pain she was obviously in, to try and run.

There was no point in leaving her to die slowly or be killed. So I drove back to the house, collected the gun and shot her. It was execution, not sport.

On examination it proved that the bone in one back leg was shattered. How she had contrived to move at all was puzzling. Worse, she was in milk and must have been nursing a litter.

I hoped the young ones were big enough to cope for themselves. But a minute or two later the clinking chain dredged up the answer. I looked behind and saw the bodies of three small leverets bouncing along with the rubbish. They had been dead a day or two, presumably because the hare had been too ill to feed them or keep them warm at night.

In the house we skinned the hare and put her into the deep freeze for future use. There was no sign of buckshot in the leg, nor anything else that suggested the huntsman. From the shattered bone I could only conclude she had been knocked down by a car. The nearest road from the fifteen-acre was half a mile, perhaps farther away. If it had been a car, it said a lot for the mothering instinct that she had managed to struggle back to her young. Every yard of the journey must have meant agonizing pain. It was pitiable that having paid so dear a price, she had been unable to save her litter.

The predatory aspect of nature was never far below the surface of farm life. Now it was spring, birds began nesting in the hedgerows, trees and throughout the farm buildings. We were blessed with a great variety of birdlife. They came in all sizes, from the grey herons which stalked the shallow streams, to the tiny red wrens which nested in the ivy blanketing the ash tree in the garden.

It seemed to us that some sort of madness possessed a few of the birds when they chose their nesting sites. One wren made her home in an old, mud-built martin's nest set on a beam over Dorfie's sty. This really was living dangerously because pigs are only too ready to accept whatever fortune may offer, even if it is simply a baby wren. Happily for them these wrens seemed to get away with it and we saw nothing to suggest that any of their brood had varied the sow's diet.

There was no such happy ending for the blue-tits who chose

to nest in the stone wall of the stockyard. They were spotted by the cats, the most dangerous predators of all where birds were concerned.

The principal villain was Fanny Fatcat, the little black moggie we had brought from London. She had been a genuine suburbanite. Initially life in the country had appalled her. Buses, trains, cars, roaring motor-cycles, townee dogs she could cope with, but the wide open spaces of the farm, the silences at night and the odd sounds that broke them, were all too much.

For the first three days she had resolutely refused to step outside the door. In the end I simply picked her up and slung her out. It had been snowing. She panicked and turned on the kitchen door. The bottom six inches of that door were sodden and rotten from rain water that had dripped on to them from a broken gutter for a great number of years. Fanny Fatcat proceeded to horrify us by scratching her way back into the house. I didn't have the heart to throw her out again: besides, we were a little worried about just how she might get back in next time.

But within a week our suburban pusscat had been transformed into a mini-tiger by the realization that the farm was really a cat's garden. From dawn to dusk she hunted rats, mice and birds, until sadly, say six months after our arrival, she picked up poison, possibly from a rat because we never used the stuff, and died. The kids buried her tearfully in the garden with her name and brief history scratched on a piece of flat limestone to mark the spot.

In the affair of the blue-tits her accomplice was one Barney the Bastard Barncat. He was a big black-and-white neutered Tom left behind by the previous occupants. He lived in the barns and arrived at the kitchen door or the dairy only for his ration of milk. In spite of this he was an affectionate old cat who loved to be petted. His affection, however, did not extend to birds.

Everything was fine until the blue-tits' eggs hatched in the nest in the stockyard wall. In due course the young birds – there must have been a dozen of them at the start – began pushing and shoving one another and crawling about the nest.

The entrance passage sloped downwards and when they reached a certain point in their explorations the baby tits simply slid down it like kids at an amusement fair, to land on a pile of builder's sand at the foot of the wall. No amused parents waited for these young arrivals, only two voracious cats who took up almost permanent residence on the sand.

The parent birds worked furiously to feed their brood and never appeared to notice that the number was diminishing, but our children did. They chased the cats away again and again but of course, the pair always returned.

Finally, after being harassed – 'Do something please, they're murdering the birds' – I found a plastic bottle, chopped the bottom off to make a crude funnel and wedged it at a 45 degree angle down into the nest entrance. It prevented the young birds sliding out, and with the supply of titbits cut off, the disgruntled cats moved elsewhere. The parent birds now had to scuttle along the funnel neck to reach their young but they were not in the least bothered. They continued to come and go exactly as before.

It was about the same time that the kids came racing up from the bottom fields to tell us about a strange bird that had taken up residence in the string of ponds which divided us from a neighbouring farm.

There it was, incongruous in its elegance, sifting the muddy water through its beak. A pink flamingo! Shirley, John and I had seen them in Africa: at Lake Nakuru, Kenya, and elsewhere. Now here was one standing gravely in a shallow pond in the shadow of the Welsh hills, while fat, red Hereford cattle browsed round it. It stayed three days and then left heading south. No doubt it had escaped from the captivity of some private zoo. We wished it well and hoped it made the tropics.

The house martins had come back at the end of April but it was two or three weeks later that their bigger relatives, the swallows, came zooming in. Suddenly they were everywhere. Summer was coming in fast.

The swallows favoured the high, secure arches of the old cowhouse. Their mud-pat homes were plastered in a score of corners and ledges in the high building, safe from cats, rats and any other creatures with a liking for eggs. Nowadays the place was used as a foodstore, although the cow ties and stalls were still in place. The roof was tiled but there were paneless windows, open eaves and a dozen or more entrances for the birds to use. Nor did they hesitate to descend to our level when it suited them. Through the open door they would come sweeping, nonchalantly lifting a wing at the last possible moment to clear an obstacle, coming close enough to startle John and myself into ducking in alarm.

Once arrived they settled into a routine of eggs and young. Like everyone else, birds, beasts and man, they had to feed their children and they were at the task all day long. As for the new generation, they lived among the remote rafters. We saw little of them until they suddenly appeared, fat and fresh-feathered, twittering excitedly on the telephone wires. But one morning I walked into the store to find the whole building alive with young swallows making their maiden flights. From ledge to beam and back again they sailed with worried parents flying among them like fighters escorting a squadron of bombers.

There were curlews on the farm too. They nested somewhere in the twelve-acre or perhaps in one of the neighbouring fields. Their plaintive, repetitive call signalled their reappearance after winter but where they had come from I did not know.

For myself, of them all I preferred the plovers and wagtails which followed the tractor when it was working on the land. The former were brave birds, transformed magically when their plumage caught the sunlight, who did not hesitate to challenge the bigger, unscrupulous birds like crows or hawks, when nests or young were threatened. They would even dive-bomb us with racketing, battering wings if we invaded their territory.

The bright, freshly laundered appearance of the wagtails always cheered me. They followed the tractor, short-legged over the upturned clods, confident of their right to be there, searching for insects or worms to feed their families.

We bought a book and became, if not ardent birdwatchers, at least interested observers of what was going on around us. Every nook and cranny around the buildings seemed to be colonized by sparrows or starlings. When I went to move an old tractor tyre in an empty calf pen, it proved to be home for a pair of very indignant robins and their three fledglings. In the fields, John, who had an eye for such things, could point to a score of nests in hedge and bush that I had not noticed. It was John who informed us that the hollow oak in a corner of the fifteen-acre field sheltered two screech owls and their brood. Sometimes, walking about with so many eyes watching, it was difficult not to feel that Egerton belonged to them and we were the intruders. So many birds and creatures were making their living from the same small plot, it seemed presumptuous to say it was ours.

20 Sheep, and a man of the soil

From our fields we looked up to the mountain, with its rough collar of small irregular fields marked with sparse blackthorn hedges or drystone walls. The mountain continued on above them, reaching up to its bluestone peak, brown with bracken in the dying months, rich in the benevolent seasons. As the light changed, so our mountain changed.

In winter it was bleak and inhospitable, like a hungry hound crouching on the ground, its flanks peppered with snow. In spring it was a jewel alive with colour, and in high summer it was a merchant's purse of velvet, greens, golds and russets flung on the rich plain.

But whatever the season, the mountain was alive with sheep. From a distance they were tiny flecks of white or minute pearls strung along the sloping tracks. Sometimes, when the collies drove them, they were like slow-moving clouds.

These sheep were mainly small Welsh ewes, nimble and hardy, able to find a living among the mosses and sparse grasses of the mountain. Tourists fed them and they were experienced beggars. If you settled down for a picnic, they would come up expectantly and did not hesitate to steal. Their lambs ran with them: fluffy, picture-book creatures.

On the lower farms we could carry heavier breeds: Cluns, Kerrys and mixtures, and run them with heavy Suffolk rams to produce the blocky, meaty lambs the butchers favoured. There were Welsh Halfbreds and Leicesters and some local products that had lost touch with their ancestors.

Everyone was agreed that we ought to 'have a few sheep', which was very reassuring because we already had some. Shortly before moving to Egerton we had had a chance, through the agent, of acquiring a small flock – sixty-five ewes – from a farmer who was moving to a bigger farm, needed the cash urgently, and was selling to start again with fresh stock. They cost us £8 a head inclusive: young, old, weak and strong, half-a-dozen different breeds. Some of our new friends sniffed at our 'bargain', but the seller was prepared to keep and lamb the flock without extra charge and that, for us, was the great attraction.

He kept his side of the agreement. His farm was about seven miles from ours, and as the lambs arrived I laboriously collected them and the ewes in Old Lil and moved them to a piece of land – ten acres – which we had been lucky enough to rent for one year. Like all lambings, there were losses and we ended with sixty ewes and eighty-five lambs; not a fantastic percentage increase but more than enough to make us happy. In addition there were two orphan lambs, twins, whose mother died soon after their birth. Locally they were known as 'caids'.

'You do bring me the nicest things,' an overworked Shirley exclaimed when I produced the frail, trembling pair. Feeding calves and chickens, organizing the family, cooking, running the children to catch the school bus and trying to find time to decorate the house, kept her trotting from one job to the next. Now I was hesitantly adding to her load.

'Never mind, give them here,' she said. Their first need was food and they sucked warm milk from a baby's bottle before being settled down in a box tucked away in a corner of the warm kitchen. That night she had to climb out of bed to feed them again at midnight. It was freezing outside and the house was chilled. She crawled back in muttering, 'Bloody lambs, bloody farm . . .'

It was my turn to get out the next night and I could appreciate how she felt. Fortunately it was not long before the pair were strong enough to get through the night on their final evening feed.

By the third day the lambs, Charles and Freda, were sprightly enough to jump in and out of their cardboard home and trot about the kitchen making nuisances of themselves. A couple of days later Shirley found them in the end lounge sleeping on her best carpet. Out they went to make their home in a tiny shed in the garden.

We also had a collie pup. She had been presented to us within weeks of our arrival. The theory was that she would be purely a working dog. None of this petting and fussing nonsense. What a hope! She was a petite, black-and-white creature, extremely intelligent and very affectionate. The kids called her Spotty Dog after a television character and taught her to beg and shake hands.

The locals did not approve and warned that she would have no stomach for work. But the instinct was very strong and when the sheep did arrive, Spotty proved very keen. It was a different thing with cattle. She had no interest in them and went through the motions of herding without any great enthusiasm.

We soon found out that although the rented land was only a couple of miles from Egerton, having the sheep there added greatly to our work schedule. They needed to be checked daily. Sheep are a strange mixture of hardiness and fecklessness. They will survive the hardest winter on an exposed mountain and then stand and unprotestingly starve to death when snared by a bramble, even though a quick tug could break them free.

On one occasion I arrived to find a small lamb with its head stuck through some old wire netting. It had given up the struggle and just stood there helplessly. When I grabbed his backlegs and yanked, he came out quite easily. There was nothing wrong and he rushed off to his mother who was waiting close by and began feeding. Afterwards they walked off together, leaving me slightly let down and feeling that someone, somewhere, should have said, 'Thank you.'

When we had the sheep at Egerton, Nicholas came puffing up one day, took my hand and said, 'Come and see. It's very sad.' It was too. A fine fat lamb had kicked himself first into exhaustion and then into death after failing to get under a wire fence although at any moment he could simply have pulled out backwards and walked away.

For anyone with sheep the biggest nightmare was stray dogs. After listening in the evenings to some of the horror stories told, I hardly dared visit our flock the next day. There was no one among our neighbours who had not lost sheep through their being worried by dogs.

Inevitably it happened to us. I arrived at the rented land early one morning to find the savaged remains of two healthy lambs. Presumably the dog or dogs responsible were lying on someone's hearthrug even as I stood there. An initial inclination to vomit at the mess gave way to impotent rage. Had the culprits been on the scene, I would cheerfully have torn them to pieces.

The locals were sympathetic but hardened to such incidents. 'Take a gun along with you,' I was told. 'If you see a dog thereabouts, let him have it first and worry about who owns him second. Best thing is to bury him quietly and say nothing.' What they advocated was common practice in the lambing season.

Stories of dogs and sheep were manifold. Holding court at a farmers' function one night, Morgan Lloyd, a pedigree sheep breeder, recalled how a sheep-chasing Alsatian had been cured. 'We dropped him into this big sort of shed with two old Clun rams,' he said in his slow, melodious voice. 'As soon as he sees sheep, in goes that dog. And he's so busy after the one, he doesn't see the other fellow coming and before you know what's happening, over he goes. He gets up damn near spitting, he's so mad, and makes for this old boy, and, of course, as you'd think, in comes the other and down he goes again. That does it. Them two boys bounced him round the shed like he was a ball. By the time we opened the door, he could hardly crawl out, but it cured him. He'd never even walk through a field if there was sheep in it, afterwards.'

It was a good story and there were others, but for every one there was a score of tales about sheep killed. In one dreadful night, not too many miles from us, a farmer lost sixty sheep. Not torn to pieces this time. They were kept flocked and packed so closely together by some dog or dogs, and the pressure of the sheep on the outside was so great, it forced those in the middle up into a mound: they suffocated.

'Mind you, that dog was no outsider,' Aaron (from up the mountain) said. 'That was a collie, and a damn good collie. Somebody forgot to tie their dog up that night and that's for certain. If it had been a townee dog, them sheep would have been scattered all over the place. There might have been one or two, even a dozen killed, but it wouldn't have been like it was.'

No one dissented from that opinion. It was a sad fact but farm dogs were all too often to blame. If anyone could point a finger at such a dog there was no reprieve. It was shot and it fell to the owner to do it. It was not a thing to be handed over to someone else.

It happened to big Geoff Bradley the first March we were at Egerton. He was in the pub one evening, slightly the worse for wear and very depressed. 'I'd to put down my old dog today,' he explained. 'Nine years old and he goes and worries my neighbour's sheep. Three of them. Never done a thing like that before. Bloody old fool.'

The whole relationship between collie and sheep was strange when you saw how aggressive a single ewe could be in defence of her lamb. They would charge and if they made contact, they could send a dog rolling. But once in a flock, fear of the predator seemed to overwhelm them. It was hard to understand why and, indeed, the whole flocking process was a puzzle.

Our own sheep would run up when we arrived to see if we had brought food. They refused to flock for a mere man. Driving them without a dog could be a wind-sapping task. You became an inferior two-legged dog and they never hesitated to double back, dodge or make a bolt for it. Nor did a short-legged dog like the little Jack Russell terrier we eventually owned seem to inspire much respect. But a collie had only to walk into the field and stand there quivering, waiting for the word to go, and ewes and lambs would run together from the furthest corners.

After our two lambs were killed it was obvious that if we were to have peace of mind, we must bring the flock to Egerton as soon as this was practical. But before this was done we needed to tail them all, castrate the ram lambs and put our identifying mark, a red E, on each and every one, ewes as well.

There was something of a ritual about tailing. 'Don't you go counting them until they'm tailed,' the locals told us. 'Don't tempt fate. And even when you've got the tails, don't go spending the money because it's a long run between the lambing and the market.'

One April morning after breakfast John and I plus two excited children and one collie pup set forth for the rented ground, equipped with a hand tool known as an elasticator borrowed from Willem the neighbour. It made it possible to castrate without cutting by slipping a small, tough rubber ring over a ram lamb's testicles so that the blood supply was cut off. Provided it was done early before the lambs developed, it appeared to be relatively painless.

Our second piece of equipment was a carving knife honed to razor sharpness for the tailing. One man held the lamb, the other stretched its tail taut and severed it with one quick slash. It was not a job to be sought after but the lambs seemed to recover very quickly. If the tails were left on they collected muck, attracted blowflies and greatly increased the risk of maggots.

From the first it was bending, back-breaking work. We started by bringing the flock into a penned-off area where they could be held and handled. The difficulty was that the pen was just a little too big. The sheep could move and did. The ewes baa-ed and butted and flocked tightly together with the lambs hiding among them. Reaching the lambs was a combination of tug-of-war and hide-and-seek. 'Like wading through a sea of sheep,' was how John put it.

In addition to tailing, castrating and marking, we had decided to dose the flock with a preparation to clear them of intestinal parasites. By the time the last lamb was done and we could lift out a hurdle and let the whole, greasy, cotton-wool mob race away, apparently determined to put maximum possible distance between them and us, we were not sorry to see them go.

We plonked down thankfully on the close-cropped turf to drink what was left of the tea in the flask Shirley had prepared. We felt rather pleased with ourselves. We stank of sheep. Our boots were thick with muck. Our hands were greasy, dirty and bloody and our nails were broken and black. John had smudged marking fluid across his forehead and in his fair hair. My spine felt as if it were about to seize up. But it was done. There were eighty-three grisly trophies in the sack on the grass. I tried to remember that it was a long time until market but all the same found myself translating the tails into cash.

That was in the future. More immediately we had to get back home and milk cows, feed calves and pigs and cope with whatever else could not be put off until tomorrow or the day after. Reluctantly I stood up and told John, 'Time to get back and start work.' The two youngsters had deserted us but the collie pup lay tiredly against my son's legs. It had been a hard day for her too.

It was almost a month later that we moved the sheep to our own land. By then they had grazed the ten acres down to something resembling a billiard table and begun trying the fences to reach the better feed in neighbouring fields.

By the locals' standards it might have been a routine operation. For us it was an adventure. The distance involved was about two miles and that, with 143 ewes and lambs, was quite far enough.

Our first task was to collect them into a flock and be sure that none was left behind. That meant scouring the gulleys and dingles. The collie pup chased about valiantly, and slowly but surely the sheep got the message and began to move. John walked in front to lead and I came behind with the pup to chase up stragglers. The lambs gave us most trouble. They started off as excitedly as children on a school outing, racing about in groups, getting lost, finding their worried mothers, and getting lost all over again. In these early stages they were bursting with energy.

Our route began along a narrow, hedged lane. Whenever we reached a gate, a stile or a gap, a batch of lambs would shoot through and chase along on the field side trying to rejoin the flock but finding no way through. If the pup went after them they panicked and scattered in all directions. Most of them eventually found their way back to the original gap and joined the ewes but every time a few had to be run down, cornered or grabbed somehow and brought back. John was young and fit, lean and hard, able to run forever, but before we were a third of the way my legs were rubber and sweat was running freely down my neck. But, never mind, we somehow completed this stage and reached the road which had to be crossed.

One consolation was that the more tired they got, the quieter they were. Holding them for the traffic to pass and give us a clear road was easier than I had anticipated. And a few minutes' rest was extremely welcome. As we waited a party of hikers – townees – stopped to chat, obviously intrigued by the drive. They asked about weather prospects for the day and, remembering the BBC forecast and ignoring John's grin, I was able to squint at the sky and tell them, 'Her don't look like raining. You ought to be all right.'

They were much impressed. The eldest of them, a man about my own age, took on himself to encourage us in our work. 'I admire you country people,' he said. 'You do an essential job producing food.'

It was very pleasing to be thought a man of the soil so I adopted my most humble forelock-touching expression and said, 'Thank you, sir. Most kind.'

I had come out without any money and rather hoped his patronage would extend to the price of a beer at the Forge which was within easy distance and could be reached if our route was changed a little. But apparently his admiration of rustics did not stretch that far. Nevertheless, it was a cheering incident for both John and me.

The last stages of the journey were along tarmacked side roads and then into our own lane. By the time we reached the farm gate ewes and lambs were very, very tired. Some of the older ewes had sore feet from the hard surfaces and limped thankfully on to the grass. Later the locals told us that walking on hard surfaces helped harden sheep's feet and reduce footrot. Whether this was true or not, men, sheep and collie pup were very relieved to have the gate closed behind them. As soon as we let them go the flock spread out over the top meadow and began feeding. John and I watched them for a few minutes and then made our way down to the house.

Shirley was reading a letter from London. 'There you are,' she said, handing it to me, 'read that. Janet and Geoff are going to Tunisia for a fortnight, the Robbies are off to Portugal. I will settle for a new floor . . .'

'And a cupboard,' Vicky said belligerently, aligning herself with her mother.

'All right,' I said. 'But could we have a cup of tea first . . .'

21 Woman-power and a new floor

Shirley had been extremely patient about the shortcomings of the farmhouse and, sometimes with an effort, had restrained herself from comparing it with the suburban home we had left. Picturesque it might be, but it also left a lot to be desired when it came to comfortable living.

In particular something would have to be done about the old brick floors in the two biggest living rooms. They must have contributed generously to the incidence of rheumatism over the years. Sometimes when I came down in the early morning, the entire ground floor smelled like a dank marsh. We tried to keep a fire burning overnight to avoid this.

But while Shirley and I might both agree 'something' needed

to be done, we differed on priorities. Money spent on floors could not be spent on stock.

'Damn your ruddy animals,' she said crossly. 'Look at this carpet. If we leave it much longer we shall need new ones and that will cost a bomb.' She turned back a corner to show how the carpet was coated underneath with mildew.

Fortunately Griff at the pub, as ever, knew someone who could help. One of his nephews was 'in the trade'.

The result was that the following Saturday two wild-looking young men – long hair, ragged jeans, old Wellingtons – arrived in a small pick-up van. The taller of them slouched over and asked in a soft Irish brogue, 'Would you be Mr Holgate ?'

'If you're selling anything, I don't want it and can't afford it,' I said hurriedly.

The second man laughed. 'I'm Griff's nephew and this is my mate Paddy. It's about a floor or something.'

Shirley appeared and looked doubtfully at them. She had expected someone older, more in keeping with the popular image of craftsmen.

Inside the house Paddy walked round the two rooms, examining corners and features. 'Nothing very difficult,' he said. 'How about £25 cash in hand and you buy the materials ?'

Shirley snapped up the deal before I could speak.

'We'll come next Friday after work and make a start. Finish Sunday night. We'll need a ton of red sand, another of sharp sand, and about eighteen bags of cement. We'll bring the plastic sheeting and a cement mixer.'

'Why sheeting ?' Shirley asked.

'Lay it down, put your concrete on top, and no damp can get through,' Paddy explained. 'No need to bother about carpets rotting then.'

They did not want tea or beer, but Griff's nephew said, 'We've got guns in the van. Can we walk round ?'

'I'll take the shotgun and go with them,' John said quickly.

When the three of them returned nearly two hours later Paddy was looking very pleased with life and carrying a big hare which he slung into the van. After they had left I asked my son, 'Who got the hare ?'

'Me,' he said. 'But don't tell anyone. Paddy is going to take it home and tell his wife he shot it.'

Shirley was like the cat who stole the cream after their visit. 'How about that, then, Farmer Giles?'

'It's a lot cheaper than I expected,' I conceded.

'So we *can* afford a cupboard in the utility room, can't we?' she asked belligerently.

It seemed advisable not to argue. The next Monday I drove into town to get the materials. Luckily the salesman at the timberyard was a frustrated DIY fiend and business was slack. I presented him with a sketch plan and he used his power tools to cut everything to size, shelves and all. Finally, because I was obviously incompetent, he showed me how to cheat on the joints by using angle irons.

As a result, when I got home it required only a few holes bored here and there in the wall and the whole thing went together like a jig-saw. Shirley was most impressed with my new dexterity. She never complained that it did not have, nor did it ever acquire, doors.

Her cup almost bubbled over at the weekend when Griff's nephew and Paddy reappeared. They worked at a mad gallop. The offending floors were ripped up and dumped outside. The bricks had simply been set in a mixture of sand and lime with the bare earth underneath.

Next morning, Saturday, they arrived as I sat down to breakfast, gulped down tea, refused anything else, and piled in for a repeat performance. It was tiring just to watch. Griff's nephew and a willing John fed the mixer and barrowed the wet concrete into Paddy, who laid the floor. No need for levels: he relied on his eye.

The biggest living room floor went in that day, the smaller one was finished by the Sunday evening. We got about the place on a bridgework of planks. A glowing Shirley would have tight-roped across Niagara had it been necessary.

'Better keep off it for about three days,' Paddy advised her as he washed his hands and pocketed the money. 'You can walk on it then but I'd give it until next weekend before putting heavy furniture back.'

Fat chance of anyone damaging her precious floors. She would have murdered them.

The pick-up was hardly up the lane when she demanded, 'Now fix the back door guttering while you're in the mood.'

'It does drip a bit,' I agreed. 'But I haven't noticed it being particularly bad.'

'It doesn't drip, it pours out,' she snapped. 'The water runs over the yard and under the door. As for you not noticing, that's not unusual, you don't notice anything but your precious farm, your precious stock and the precious cost. You wouldn't notice if I went bald.' She managed to work herself up into quite an impressive state.

Fortunately Thomas came up with a message from his father-in-law, Ellis the cowman, and pointed to a simple solution. The water poured out because the guttering sloped towards the back door and there was no downpipe or stop-end, or anything, to hold it back or lead it away. Furthermore it had been doing so long enough to wear a hole two inches deep in the concrete of the yard. 'We can lift it up so that the water runs in the other direction, away from the door,' he said. 'We'll put in a down-pipe, lay in a drain to the field ditch and let the water run away.'

He came up for a couple of days. I laid out a few shillings on a U-bend pipe for the drain and we found the other bits and pieces round the farm. It ended a nuisance that must have made life unpleasant for several decades of housewives.

Shirley was appeased for the time being but took all the credit for herself. She went round convinced that without her nagging nothing would have been done and muttering about 'woman power'. She might have been right.

22 Shirley's amazing parsnip wine

'Maison Egerton' owed its origin to a slim book presented to my wife by one of our city friends. The author reduced the wine-making operation to the barest essentials and scrapped the mystique. The method seemed to consist of slinging whatever fruit or vegetables could be obtained into a plastic bucket, adding a little yeast, topping up with water, and leaving it to ferment.

In her role of farmer's wife, Shirley, looking our friends said, more like Vicky's elder sister than her mother, became extremely enthusiastic. Dressed in a green smock which was a relic of her schooldays, book in hand, she lectured the family on how to use the most unlikely ingredients to produce wine guaranteed to have Frenchmen tearing their hair in envy. The cupboards began to fill with an assortment of bottles and containers which burped and gurgled like alcoholic uncles. Not one corner of the ground floor was without its resident bucket. The end result was not always up to expectations but beggars cannot be choosers and, ultimately, whatever was produced was consumed.

One of the biggest difficulties was that according to the book, the wine should be allowed something upwards of six months to mature. With the exception of her potato plonk which really was dreadful, none of ours had any chance of even approaching that time, particularly since our London friends kept arriving for weekends in the country. They acted as guinea pigs and were too polite to express their real opinions. We plied them with Maison Egerton and unscrupulously saved the French wines they brought with them to drink after their departure.

In her career as a wine-maker, Shirley achieved two notable successes. The first came early on and helped establish her reputation among local wine-making circles.

It began when I arrived home from market with a sack of

parsnips which had been going too cheap to miss. There were too many to eat, so the rest were fed into Maison Egerton to be converted into wine. We discovered that among our neighbours parsnip wine had a reputation for being alcoholic. Shirley's product made that the understatement of the year.

Most of the locals – and Howard and Ellis were no exceptions – fancied themselves connoisseurs of farm wines. Both were Shirley fans and intrigued with her efforts. But this evening there was just a hint of condescension in Howard's tone when he said, 'You ought to try Dilys's parsnip wine. I've never tasted anything to touch it.'

'Really?' Shirley asked. 'Would you like to try mine and see how it compares?' There was something in her tone that made me uncomfortable. However, she went off and returned with a brown jug, half-filled with the stuff and two glasses.

'It hasn't been made long,' I explained hurriedly as she poured out generous tots. 'Perhaps it's too young to be judged properly.'

In the manner of experts everywhere, Howard held up his glass to the light. 'Doesn't look too bad considering it's had no time at all to stand. It always improves with the keeping. Dilys won't touch hers for a year.' Ellis had similar words to utter.

Both drank and both gasped. 'Holy cow,' Howard gasped, with his eyes watering. 'I've never tasted parsnip wine quite like this. What's in it?'

'Oh,' she said sweetly. 'Don't you like it? It's only parsnips, sugar, yeast, all wholesome things, and a little something of my own. Perhaps it was just the first taste. Drink up and try another glass.'

The second glass confirmed their impressions of the first. Both looked decidedly happier but slightly wobbly. They allowed her to pour out a third sample but stopped her hand before she could fill their glasses.

'Must be something about the parsnips,' Ellis said more thickly than I had known him speak. 'It warms you right through.'

He was beginning to look very hot indeed.

'I thought you might find it interesting,' she said demurely. 'I'm glad you like it.'

'You want to have a few words with Dilys,' Howard said, wiping his eyes and getting up from his chair with a conscious effort.

'Why?' she asked, all innocence. 'Doesn't Dilys's parsnip wine taste like this?'

Our guest steadied himself. 'Not quite the same, perhaps. Never known parsnip wine taste quite like this.'

'I must have been lucky then,' she told him; she was up to something all right.

They left us singing her praises and set out, rather unsteadily but very merrily, to walk to the Forge where Thomas was waiting with his car.

'Let's see our precious Dilys follow that,' Shirley said triumphantly, as we watched them leave.

'Well, wife,' I said. 'Now that we are alone, tell me the secret of your success.'

'Trade secrets,' she said airily. 'Nothing a mere man could possibly understand.' Nor would she go further, and it was a couple of weeks before I realized that half a bottle of vodka left over from some suburban Christmas had vanished.

Her wine-making endeavours throughout the year landed us with an assortment of plonk, all of it drinkable, some of it enjoyable, made from just about everything that grew in or on the ground, or hung from bush or tree. The kids scoured the fields and hedgerows on her behalf. There was dandelion, apple, blackberry, rhubarb, plum, elderflower and – my favourite – elderberry, which had a very pleasant taste and appearance. One of her most potent brews was made from the bitter sloes which were common round the farm and, surprisingly, resulted in an almost too sweet but extremely potent wine.

But she really had to wait until the cherries ripened to notch up a near-repeat of the parsnip triumph.

There were two trees, loaded with fruit, one luscious Whiteheart, the other a smaller, bitter red fruit. Nicholas Paul and Victoria Jane plus visiting juveniles ate themselves silly on the Whitehearts, and Shirley put bowls of them on the tea table and muttered vaguely about preserving without actually doing anything.

Hilariously, the hens proved among the most appreciative cherry fanciers. They put on a tremendous display of gymnastics, jumping vertically into the air and attaining fantastic heights to snatch low-hanging fruit which they swallowed, stones and all, in one throat-bulging gulp. If they landed with a cherry in their beak they were immediately mugged by their waiting companions. One of them, somehow, broke a leg and was found floundering on the lawn, so into the pot it went. Its crop proved to be so tightly packed with cherry stones there was little, if any, room for anything else edible.

In this cherry-orgy no one fancied the smaller, red fruit, but rather than see them waste, Shirley picked a bucketful and made wine. It was sweet, smelled like cherry brandy and packed a kick like a mule. One generous glass was sufficient to ensure that our visitors saw farm life through a rosy glow. Several of them could not remember seeing it at all. One very proper lady stunned us by breaking into, and insisting upon singing, a very rude rugby song.

There was no shortage of volunteers to pick the following season's crop but, alas, there was hardly a cherry to be seen.

Shirley's wine-making efforts sparked off a fashion among our suburban friends. They appeared on the farm laden with jars, cans and all sorts of containers and wandered round the place picking a variety of berries. It was nice to see them, they kept us in touch with the life we had left, aspects of which we often missed. Mainly, I think, it was the bondage that farm life imposes that irked in those first tiring months. Shirley felt it more because I had been accustomed to the treadmill of commuter train and office routine. But both of us at times felt the need to get right away from it all, if only for an evening and this, unfortunately, was extremely difficult to do.

 23 Taffy the urchin calf 

One of the pleasantest aspects of farming was that the results of one's toils were easily visible. It was gratifying to be able to walk round the fields and see how the stock was increasing, even if the bank overdraft tended to show a corresponding increase. By now we considered ourselves well on the way to becoming farmers.

With the better weather the kids rambled all over the fields. They had special names for many of Egerton's features. A favourite place was where a spring bubbled up in one of the fields and emerged from the tangled green roots of a clump of holly as a tiny, trickling stream. They named the spot 'Miracle' and spent long, happy hours building mud dams to collect the water and make a drinking pool for the animals. A major constructional hazard was thirsty cows who waded into the reservoir and demolished dam walls with one touch of a dinner plate hoof. A corner in the fifteen-acre which had been left uncultivated and was rough with bushes and tall, rank grasses was 'Pooh Corner'. They had a score of hiding places – it seemed unnecessary to have anyone to hide from – and special climbing trees graded according to the challenge they presented.

Willem's younger son tagged along with them; a strong, dour lad who listened stolidly to their prattlings but was very ready to join in and play out their fantasies.

For we three bigger ones there was time to stop, look round and take note. After all the hurryings and scurryings to get the place functioning, a quieter period was welcome. It was on one still evening in this phase that Shirley and I were walking round our tiny kingdom when she said, 'Know what? I'd like to put on an evening skirt and go somewhere special for a meal.'

'Like where?'

'Soho, Dean Street, Bertorelli's, somewhere like that.'

'It's impossible,' I said. 'We can't get there and we couldn't afford it if we could, but there is a market tomorrow, how about that?'

'Not exactly what I had in mind,' she sighed, 'but it will have to do.'

We had nothing to sell but every market had its own fascination. It began when you arrived, one of hundreds of farmers converging on the spot. There was a queue for the car park. A big cattle lorry had found he did not have the necessary lock and was being forced to edge his way in.

This time we went in the 1800, not the old wine van. As we waited, a small pick-up pulled alongside and a dark-haired, plump woman sitting with her husband wound down the window and asked, 'What's going on?'

'Nothing serious. Just a lorry stuck in the entrance.'

Her husband said something and she asked, 'What time do they stop taking entries for sheep?'

'You've got a good half hour yet,' I reassured her. 'Bringing lambs?'

'No,' she jerked her thumb to indicate the back of the vehicle. 'Just an old ewe. The only one in the flock that didn't get a lamb. No use keeping her.'

The jam began to clear and I held back to let them go ahead. As they pulled away, the barren sheep peered out of the wire-netted back doors. She was grey-muzzled and hollow-cheeked. In her life she must have produced many good fat lambs. Now her reproductive abilities were exhausted and she was to be sold as a slaughter ewe, destined to become processed mutton. It was sad but I suppose in a natural environment the four-legged hunters would have pulled her down a long time before now.

This day the main trade was in Easter lambs. They were still called this although the holiday was well past. In the main they had been born in January and were being sold now before the main crop of lambs came to the market. They were in short supply and fetched good prices.

A market is a machine. The process began here with the lorries and vans bringing the lambs to the reception point. The vehicles pulled up against the unloading ramps one after another and the lambs were off-loaded and ran in batches, usually about eight or ten, on to a weighing platform. Their gross weight was called out to a clerk who converted it to an average 'killing out' weight which was what the lambs represented in marketable carcases. Those on offer varied between 32 lb and 46 lb.

From the weighing point the lambs were taken through the aisle to the selling pens. When selling began the auctioneer and his entourage strutted along a narrow catwalk which ran above the pens; he sold by reference to carcase weights, singing out the bids, wisecracking with purchasing butchers and selling farmers, bringing each transaction to an end by thumping his heavy walking stick on the pen rail and chanting, 'Sold to Mr . . .'

His clerk entered the figures on a pad, and as the pages were completed they were torn off and carried to the office staff who finalized the transactions, collecting cheques from buyers, arranging credit, paying out to sellers, calculating commission fractions, counting five-pound notes and one-pound notes, writing cheques, or, most commonly, paying out in combinations of cheques and cash.

By chance we saw the old ewe sold. There were a couple of dozen like her on offer. No one wanted them much except a young Asian who got her and two others in the same pen for about five pounds each. According to a farmer near us, they were intended for the 'Pakistani restaurant trade' but whether he had grounds for so saying or whether it was just speculation, I did not know.

Shirley wanted to look at the market garden produce. It meant passing through the pig market. Selling had been completed and the pigs were being loaded into lorries, many of them double-decked, to be taken to the abattoirs. In the main they were porkers weighing between five and eight score pounds and reared to produce the lean pork favoured by city housewives.

They were hustled in batches along the aisles separating the pens and then up the sloping tailboards into the lorries. To keep them moving the drovers used electric prods, and these applied to porky backsides produced high-screeched protests and sometimes resulted in those behind trying to climb over the backs of those in front.

A white-coated butcher watching the loading shook his head in disapproval. 'They shouldn't be used,' he told us, indicating the prods. It seemed a rather strange sentiment for a man in his business. 'It only gives them a short, sharp sting,' I volunteered. 'I wouldn't think they are hurt much.'

He regarded me as if I were sub-normal. 'It leaves marks on the meat. You buy them that's been prodded and sure as that you'll find two little blood spots where they've been touched. The housewives don't like that at all. It takes away from the appearance of the meat.'

'Oh,' I said. There seemed little else that could be said. 'I didn't know that.'

He took a florin from his pocket and pointed out the milled edge. 'See this? If you were to roll that on a pig's back, when you went round to the abattoir to collect it the next day you'd see the mark of it. I've done it myself. Behind the ear, though, not anywhere it shows, just to check that I've got the right carcase. They'll swop carcases sometimes if they get a chance. But you use a coin, you can be sure of finding the pig you sent in. You remember that.'

I promised to do so and thanked him for the tip.

The produce section was the smallholders' and gardeners' show place. There was a wide variety of vegetables and fruit, chickens, ducks and geese in wire cages, eggs, farm butter, racks of rabbits and hares, the latter with their heads in tiny buckets, wood pigeons in bunches like feathered blue-grey grapes, poultry dressed and ready for the oven, and anything else that country folk thought might bring in a few pence.

In an outer room there was a collection of secondhand furniture, farm implements, tools, old books, a battered bicycle, jumble, two boxes of sinuous ferrets, and items that could only be described as rubbish.

We picked our way through the offerings but found nothing of interest, and after watching some worn-out battery hens with rumps picked red raw and bare of feathers by their pen companions sold for twenty pence apiece, we moved on to the calf ring.

This was one of the star attractions of the market. The auctioneer was already in his Punch and Judy box shouting the bidding in a hoarse, strained voice and completing each sale by cracking his gavel on the planks in front of him.

The small amphitheatre with its tiered concrete stands was three parts filled with prospective buyers and mere spectators, with more arriving as the lamb sale progressed and released men to see whether it was worth laying out some of the cash they had just made on a calf or, perhaps, two.

In quick succession the calves, most of them about two weeks old, were ushered in, bawling and protesting, and walked round the strawed ring. In the main they were sturdy, white-faced Hereford-Friesian crosses brought in by local dairy farmers. Prices were high. Depending on looks and stature they averaged between £40 and £45. Some, usually black bull calves, went higher and some, of course, went considerably cheaper.

One handsome bull calf bounded into the ring like a Jack-in-the-Box, threw back his head and roared for the cow he had been taken from that morning. One of the dealers who hogged the ring rails immediately offered £50 and after a flurry of quick bids the calf was knocked down to him for £65.

Shirley, as ever, was charmed by the young animals. They were pampered, even if presently bewildered animals. Their coats were glossy with colostrum and their neat hooves were black and shiny. When they stretched out their necks and bellowed their outrage, it was with the assurance that they were entitled to better treatment.

Would-be buyers marked their points and looked hard for any tell-tale traces of scour – calf diarrhoea – round the tail. When a particularly good calf came into the ring there was a stir and murmur of interest.

But then, in the midst of this rich stream, a tiny, Jersey bull calf was walked in. By comparison with his predecessors he was fragile and he looked miserable and neglected. His ribs stuck out and his stomach was empty and tight against his backbone. No colostrum and precious little milk had been wasted on him. He moved unsteadily round the ring and was eventually knocked down to a butcher to become veal for thirty shillings.

'Poor little thing,' Shirley said, shocked. 'Why doesn't anyone want him?'

A thickset, red-faced farmer answered for me. 'Not worth your money, Missus. He'd take as much work and near as much grub to raise as a good'un and at the end of it, what have you got? Something the butchers don't want. No. If you'm to spend money, spend it where it'll do most good and bring most back. Why buy trouble?'

There was no arguing with that advice.

Nevertheless, when a calf that looked to be something between a Teddy Bear and a scrubbing brush was brought in a few minutes later and also seemed doomed for veal, she asked, 'Why don't we get him?'

'Not worth much,' I said with all the assurance of my long farming experience.

Shirley persisted. 'He's not a Jersey, though, is he?'

'No, Missus,' the thickset farmer said grinning. 'There's a bit of Aberdeen or Welsh Black there somewheres. Lots of other bits too. But he might be worth a couple of pounds. There's a bit of age about him, have a go if you'd the fancy.'

There was no arguing against the pair of them and so I raised my finger twice and ended with 'Taffy Calf' being knocked down to me for the princely sum of seven pounds. Our thickset friend was highly amused.

When we went to collect our purchase from the pens at the rear of the building, two tow-haired teenagers were saying goodbye to him. They were relieved he had not gone for veal. It appeared they had been given 'Taffy' by a farmer they had done some work for and had kept him in a garden shed and fed him powdered milk. They hunted around, found some scrap sacking and binder twine and helped me to wrap his legs and get him into the back seat of the 1800.

The thickset one turned up to see the fun and lend a hand if needed but Taffy went in without much fuss. He was so used to being handled and mauled about that nothing appeared to upset him.

'Likely he'll drive if you let him,' our farmer friend said. 'Leastways you shouldn't have much trouble getting him settled down.'

That proved a safe prophecy. For the twelve-mile journey home he was content to lie on the backseat and suck Shirley's fingers when they were offered. On arrival at the farm he came out easily and once the sacking had been removed he trotted down the yard to join Ferdinand's mob as if he had always lived there. Half an hour later he was feeding greedily from a bucket. It was very obvious that rearing this calf should not be too difficult because almost anything must be an improvement on what he had previously known.

He was very different from his pen companions. A street urchin among the pampered well-to-do. The rough coat gave him a tatty, neglected look, and proximity had robbed him of any reservations about people and made him aggressive in his demands for attention. If there was food about, Taffy could be counted on to head the queue. If there was a finger to suck or a hand to nuzzle, he was quickly there. He loved playing with the kids and he and Ferdinand – with whom he formed an enduring friendship – would race round the pen, kicking up their heels in sheer high spirits. In time his rough coat gave way to a handsome black one but his character changed not one jot. It was a case of once an urchin, always an urchin.

24 A plague of hikers

Summer went on and on that year. The second week in June a cuckoo came sailing over the farm with his long-tailed, dipping flight and landed in the big ash tree. From there his soft, sweet call echoed over the fields. It was as if he had charmed the sun into coming to live with us.

It blazed over the hayfields drawing up the grass. It dried out the winter mud and made concrete of the lane. It was easier to climb out of bed in the morning when the alarm sang. The world was altogether a balmier place.

Had they paused to consider the point the dairy cows would

undoubtedly have agreed. Each morning they lay chewing the cud, reluctant to be disturbed by the biped who appeared, waved his arms about and shouted hopefully, 'Come on, cows.'

One or two would stir, even climb to their feet and stretch and defecate fluidly and noisily, but the majority stayed exactly where they were, made lazy by sweet grass and warm nights. There was nothing for it but to walk round and put a foot against their rumps. Then, and only then, would they get up and deign to plod towards the milking parlour.

On such days the living was good. We were putting out over forty gallons of milk daily and our monthly pay cheque for it was approaching £180. If we were not prospering, we were at least surviving. All was well with our little world.

Nor were we alone in thinking the world a pleasant place. One day, walking across a field to inspect a boundary fence with a neighbour, he forgot himself so far as to exclaim, 'This is how God meant men to live. Not in them damn smoky, noisy towns.'

He was embarrassed by the outburst – in that society it was not done for men to express themselves in such a way – and reverted immediately to his normal reserved silence.

But the sun that chased winter from our bones also brought a threat to our tranquillity. With it came the intruders: the strollers and, far worse, the hikers. The former name we gave to the comfortable, pudgy middle-ageds who, usually on Sundays, brought their cars as far down the lane as possible, parked them on the verges and ambled about in a leisurely way with their pot-bellied dogs. They seldom did any real harm.

No, the main threat came from the 'hikers', the shock troops of aggressive young and not-so-young people who arrived with maps and charts and a determination to rescue the countryside from the wicked farmers.

'Where do they go?' I asked Willem, referring to the erratic stream that moved towards his small farm.

'Any damn place they think,' he answered cryptically, and it was the stark truth.

The temerity of some of them was something to behold. Properly organized parties were no trouble. They, at least, knew where they were going, where the footpaths ran and where they could walk. They could appreciate too that the land they walked over was someone else's factory floor, that some character had been toiling over it all winter, and that nothing was lost if they walked round a crop rather than ploughing right through it.

Not so the shock troops. They were dedicated to keeping the countryside open for the people and they would brook no argument and heed no advice. Thus one day, after early morning rain, three heavily rucksacked girls presented themselves in the stockyard and demanded rudely that we inform them where the public right-of-way ran. By the map it went right through a field of growing hay. They would be better advised to go round two sides of the field to reach the far gate. Not them: no sir. No cunning farmer was going to realign a footpath in such a way. 'We prefer to exercise our rights to follow the footpath,' the tallest, a physically attractive brunette, told us haughtily. And off they went.

The grass was high. Within twenty yards their trousered legs were soaked and their boots were sodden and clodded. But they persevered and half an hour later we spotted them, lost, floundering across a neighbour's barley field, picking up mud with every stride and heading for a high, barbed wire fence.

Even they were better than the young men who, for some strange reason, kicked Willem's gate into matchwood when taking a short cut to the road. Why they didn't just climb over it I could never understand.

Another bunch found a local farmer's tractor parked in a field ready for the next day's work. They tried to start it, failed, took the brake off and allowed it to run downhill and overturn in the ditch.

Happily, it was not all 'war' and we were provided with some lighter moments. There were the seven lady hikers who strode into the dairy, helped themselves to water from the tap without acknowledging our presence, and then strode away, still as purposefully, across the top field. Now it so happened that the cows were grazing that field and when Miss Arabella saw them she decided to walk over and introduce herself. They did not reciprocate. When they saw her coming the stout, no-nonsense, booted leader waved her arms and shouted, 'Scat! Go on, scat!'

No one had ever shouted 'Scat' at Miss Arabella before, so she decided, something very special must be happening. It was worth investigating because, after all, a dairy cow's life is not exactly packed with excitement. She began to trot.

This brought an immediate reaction. The stout one yelled 'Bull!' and headed for the gate on the far side of the field. The others followed and Miss Arabella found herself in danger of losing all these interesting people. She shifted gear into a shambling, rolling gait which, with everything flying about and wobbling, was quite something to behold. The withdrawal became a full-scale rout. It was every woman for herself, and the devil take your sister.

Even so, considering she started from scratch and was not a racing cow, Miss Arabella was not disgraced. She overtook and passed three of the women before reaching the gate and peering over it expectantly, still baffled as to what it was all about.

The overtaken trio ran down the hedge and forced their way through bush and bramble at a weak spot to join their gasping, gesticulating companions. For her part, the disappointed cow hung round the gate reluctant to accept that there was nothing to be seen. Eventually, however, she conceded the point, gave a cow's equivalent of a shrug and went back to her companions who were just coming up to see what all the excitement was about.

On a more pleasant occasion I discovered a young school master and his wife with about a score of children enjoying their sandwiches while seated happily in one corner of a hay field, having first carefully flattened the grass. They were, he explained, from a Black Country school and on a walking tour to learn something about the countryside at first hand. For some of the children it was the first time they had walked in 'real' fields as against municipal parks. It seemed an admirable thing to me and I had not the heart to complain about the hay.

Hikers, however, were not the main threat to our precious hay crop. Everything that ran about the farm on four legs, except the dogs and cats, cast covetous eyes on it. Fences and hedges which had looked insurmountable barriers in winter suddenly seemed as full of holes as a length of lace.

The sheep were the main villains. Making it worse was the fact that they were heavy in wool and would soon need to be sheared. The best price was given for fleeces taken off in one piece and we needed all the pennies we could get. Every cluster of wool hanging on the barbed wire caused an ache in our pockets.

This was nothing to the flock. What concerned them was the knowledge that new grass as tender as asparagus tips was growing just the other side of the hedge. Led by a couple of old ewes, they patrolled the boundaries like ghouls. The slightest suspicion of weakness and through they went like woolly tanks forcing their way in, ripping their precious fleeces to shreds.

Once through, they fanned out on to the grass to feed. They spoiled more than they ate. If anyone appeared they rushed pell-mell, full of guilt, for the hole they had made, but they never went far and were soon back looking for another way in.

'Don't you go fretting, working yourself up to a frazzle,' Old Jonathon said. 'You find that leader and yoke her. Yoke the lot if it's necessary. You just watch 'em, though, you'll see which uns needs it.'

He was right. No difficulty identifying the hedgebreaker-in-chief. She was as bare of wool as a newly clipped poodle. We cornered her and tied three poles in a triangle about her neck. She could graze but when she tried to force her way through the hedge, the yoke caught and held her back. It did the trick. I think she knew what a yoke was from past experience. There were one or two further attempts to break in but they were half-hearted compared with previous efforts and the hay was safe.

25 Shearing, the young men's game

The cuckoo sun which drew up the grass also caused the grease to rise in the sheep's fleeces. They needed to be sheared. They were hot and uncomfortable and in danger from the big blow-flies now beginning to appear in numbers. These flies laid their eggs in the thick wool, mainly round the sheep's rumps, and the maggots that hatched out fed on the animals' flesh, sometimes with fatal results.

We registered with the Wool Marketing Board and their agent came in a small van to present us with a kit consisting of an addressed postcard to be mailed when the wool was ready for collection, labels and three 'sheets' which were mammoth sacks.

'Shearing,' Howard said, 'is like chasing girls. It's a job for young men. It's no good getting an auld mon, they might know what needs to be done but they can't get down to it, or if they can, they can't get back up again.'

It was his way of saying that shearing was a skilled craft and extremely physically demanding. For the local young men it was a chance to make money – although they had to earn it. Some local flocks ran into hundreds and a contract to shear them at 12p a sheep represented a fair sum.

The shearers were proud of their ability. It was one of those operations where the end product all too obviously showed the skill of the workman's hand. As Matthew – Old Jonathon's brother – said, when looking at a badly sheared ewe, 'Her looks like some poor bloke that's had his hair cut with a blunt kitchen knife.'

In his younger days Griff the publican had been one of the district's best shearers. 'Not now, though,' he said. 'You've got to have a flexible back. Mine's like an old stick that'll break before it'll bend.' He fixed us up with a slim young fellow, Morris Jones, who promised to 'fit us in' between bigger flocks.

Shearing calls for a clean, hard surface to work on, and on which to wrap and pack the fleeces. Foreign bodies like bits of muck or straw in the wool reduce the price. So we closed off a section of the collecting yard and swept, scraped and finally washed and broomed the concrete in readiness.

Jones phoned and gave us a date, and the night before shearing he arrived and prepared his tackle. It was all electric and he hung it from a cross beam to get all-round movement. We found a wooden platform in an outhouse and dragged it in for him to stand and work on.

Next morning was a mad scramble to get the milking and feeding finished before he arrived. The flock had to be brought in. I bolted down breakfast while he checked his equipment and put on overshoes made of sacking to give a better foothold when the floor became greasy which, he assured us, was bound to happen.

'Let's have the first un then,' he said when all was to his liking; and we were in business.

John and I caught the ewes and brought them to the working point. Many of them were as heavy or nearly as heavy as me and heavier than my tall, slim son who weighed less than ten stone: it was easier said than done, particularly since they were not prepared to cooperate.

Once on the platform they had to be upended and sat on their rumps ready for clipping. Where we struggled inexpertly with them, Jones needed only the minimum effort. They lay quietly against his legs, more like stuffed toys than living animals, as he worked. If the clippers did nip them, he daubed the spot with an antiseptic soothing cream, but it seldom happened.

There are, he told us, a number of shearing systems, but he preferred a New Zealand method. All the clipper strokes were known as 'blows' and delivered in a set pattern which took off the fleece with the least possible effort and movement. First to be sheared was the belly wool. It came off separately. The remainder of the fleece came off in one piece.

John and I rolled the fleeces into bundles, secured them by twisting the tail wool into a rope, winding it round and tucking it under itself. The bundles went into the waiting sheets.

The electric motor whirred away and Jones, who seemed blessed with an elastic spine, bent to the task, hardly straightening up to exchange the shorn ewe for the one we had waiting. The stripped sheep, looking undressed and rather ridiculous, were daubed with the identifying 'E' for Egerton and ushered back into the flock by the collie pup. The first few ewes sheared seemed to vanish among those waiting their turn, but gradually the sheets began to fill with fleeces and the point approached when the shorn sheep began to outnumber the unshorn.

We were busy with the broom to keep the area clean, and with it all were glad of a break when Shirley arrived with a jug of tea and sandwiches.

'Nothing to eat for me,' our shearer said. 'The less you have in your stomach when you'm doing this job, the better. Eat a lot and you'll end up with the cramps.'

He was a pleasant young man who had spent a year at farm college. Land prices made it unlikely that he could ever buy his own farm, so his ambition was to manage one for someone else. His family lived on a small estate holding. 'But that'll likely be taken back when the old man retires or dies,' he said. 'Mind you, I wouldn't mind working for his Lordship. A job on the estate, say shepherd, would do me fine.'

The work started again and we brought a big, old ewe to him.

'Smell them?' he asked suddenly, straightening up. 'Maggots. Can you smell them?'

There was a faint but discernible, sweet odour. He separated the thick wool about the ewe's tail. There was a whole, seething colony of them. The clippers cleared away the wool to expose the area underneath red and raw, but no lasting damage had been done.

Jones cleaned them out and rubbed machine oil into the wool about the spot. 'That should kill any that's left,' he explained. 'There's no real harm here but you've got to keep a look out in this weather. You want to get 'em round to Old Jonathon's and run them through the dip. That'll keep the flies off.'

We had already spoken with Old Jonathon, but there was a queue and we were well back in it. Throughout the hot weather we checked periodically for maggots. The worst we found was a fat lamb which appeared to be spitting out maggots. It was biting them from a patch on its own rump. We cleaned them out and used a carbolic solution to kill any infection and finish off survivors. The wool grew again, but it was darker than the original and the patch always made it easy to single out that particular lamb.

Just after noon, the last ewe was sheared. Two of the sheets were bulging with fleeces, the third had not been needed. It was a relief to be able to turn the new-look flock back into the field. They looked much more comfortable. The weather forecast was good and there was little danger of losing any of the newly shorn through sudden drops in temperature at night which sometimes happened.

The three of us scrubbed ourselves clean of grease and muck and went in to eat. Jones ate little because he was moving on to another farm that afternoon. Before the short season ended he would have lost a stone or more in weight.

After he had departed with my cheque in his pocket, we cleared up. The sheets were laced shut and labelled and manhandled into the store to wait collection. Shirley posted the card the following morning; a fortnight later the agent's lorry appeared and our wool went off to be graded and priced.

The reward for all our effort came some three months later in the form of a cheque for £63 – slightly over £1 a fleece. Not riches, perhaps, but very welcome because our sheep were primarily meat animals and wool was only a secondary product.

26 Dipping the sheep

It was another bright, burning day when we walked the flock over to Old Jonathon's farm to run them through his sheep dip. It had been discussed previously but left for him to say when it could be done. More established members of the community had prior claims. But at last he telephoned and, after paying his respects to Shirley, told her, 'Tell your husband he can bring the sheep tomorrow, if the weather stays fine.'

It did. So after a mammoth breakfast, I set out with John, Vicky and the collie pup. The distance was a little under two miles, up the lane and along a steep-banked country road to the 200-acre farm. It had been bought in the 1920s for less than £3,000; a fact Old Jonathon never wearied of repeating. His story always ended, 'Ah, but what a job it was to raise that amount of money in them days.'

Company was life itself to Old Jonathon. Friend or stranger was greeted with the same enthusiasm, and children like Vicky were doubly welcome. He was waiting for us. He indicated the open gate and the collie pup raced round and turned the flock into the yard. She was developing an anxiety about her charges which amounted almost to a neurosis. Now she set to patrolling the fences with a dedication which amazed the big, floppy-eared foxhound pup Old Jonathon was 'walking' for the local hunt. When the hound would not take a snarl as a hint but insisted on joining in what to him must have seemed an excellent game, she turned on him viciously and sent him stumbling and backing away as much in shock as hurt.

'By God, her's got some guts,' the old man's outsize employee, Billy, said admiringly. (He had arranged his own work to lend us a hand.) 'That's taught him a lesson or two and her not half his size.'

In the end we had to tie her to a post to stop her interfering. She cried and whined but finally accepted the position and lay down, head on paws, to watch.

Old Jonathon's brother Matthew arrived, limping from his arthritis, and set up a little portable, diesel-engined, shearing machine to trim any sheep's tails that might be burred or clinkered. One or two of them had green muck round their rumps. 'That's a sign of gut worms,' he said. 'You wants to dose 'em when you gets a chance.'

There was little the pair did not know about sheep. They were noted breeders. When their young stock came up at local sales it was snapped up quickly by knowledgeable farmers.

'Well, now, you ought to start getting some of these lambs off to market, Jacky,' Matthew said, a frown of concentration on his broad forehead. 'There's a few here well up to eighty pounds, if I'm any judge.'

Old Jonathon, almost lost in a pair of PVC leggings which protected his trousers, was stirring something into the oily dip solution. 'Get rid of them, young man,' he advised. 'When stock's ready, it pays to cash it before everybody and his uncle starts bringing their lambs in and the price comes down.' He looked at our lambs. 'You've got some good old ewes, I'd say. You can tell by these lambs. Old ewes always bring them on that bit quicker. But don't you get too many old uns in the flock. You keep putting in one or two younger animals. Always add to one end of the flock and take the old ones away at the other end.'

We began working. John stood to one side armed with a long, wooden paddle to duck any sheep that did not go under with the initial shove. The important ingredient of the solution was an insecticide which killed unpleasant pests like sheep scab, blowflies and maggots, lice and ticks, and also protected against reinfection.

I hung back and watched as Billy grabbed a ewe and put her into the narrow, rectangular bath. There was hardly a ripple. Under she went, bobbed up again and scrambled, dripping, out at the far end. The sloping landing ramp where she emerged meant that the dip draining off her ran back into the bath.

Much encouraged, I caught hold of a big, bony ewe and began shoving her towards the bath. All hell broke loose. She dug her front feet in stubbornly and refused to budge. I seized her round the middle and succeeded in lifting her back feet off the ground. Not one yard forward did she go. In the struggle I lost my footing and ended sitting on the wet, mucky ground.

'Keep to her,' Old Jonathon roared, delighted at the commotion.

My reputation was at stake. It was me or her. I managed, somehow, to drag the wretched creature to the edge and hurl her in. There was an almighty splash. I was soaked with the foul-looking dip and so was John.

'Now you'm both good against maggot,' Old Jonathon roared.

Billy took pity on me. 'Look,' he said quietly. He caught another ewe, turned her round and backed her easily into the tank. 'Arse first, that's what does it. Never let them see what's coming.'

My next effort was far better. Besides, I quietly determined to allow Billy to handle the bigger ewes and to content myself with the smaller animals and the lambs. It was a wise decision. The sheep began to go through at a fair pace; they emerged at the far end in a regular stream.

About the halfway stage Old Jonathon called a halt and disappeared into the farmhouse to reappear with a bar of chocolate for Vicky, a great stone jar of cider and a horn cup.

He was proud of his cider. Every year he bought about two tons of apples and turned them into a clear, dry drink which had gained a certain reputation in the area. I had been warned against over-indulgence. 'Try this,' he said and poured a liberal tot into the cup. 'This will take the taste of dip out of your mouth.'

One swallow was sufficient to convince me that it would indeed – or if it didn't remove it entirely, it would at least make it more acceptable.

We leaned against the wooden rails and talked of things farming and passed round the horn until I stayed his hand and said, 'That's enough for me, Jonathon, or I'll end up in the dip with the sheep.'

'Do you good,' he said. 'Keep the flies off.' But he did not press the drink and soon we were working again.

Perhaps the cider did help. The ewes certainly seemed lighter and more docile. In a short time there were only a few quick lambs to catch and put through. Then the flock, damp and brown from the solution, was run from one pen to another via a narrow aisle for an agreed count. The charge was two new pence a sheep and I settled up with Old Jonathon in his office, which also happened to be his greenhouse and the barber's shop where he sheared his friends.

'We'll wait for it,' he said. 'You'm welcome to pay when they'm sold.' It was an appreciation of the tightness of money round our house, but we had an understanding bank manager, so I thanked him for the thought and wrote the cheque.

The stone jar came out again and we had another couple of rounds and one for the road. Old Jonathon picked a bunch of flowers for Victoria Jane to take to Shirley with his compliments. He also sent her the offer of the gooseberries on a large, overgrown bush in his garden if she cared to pick them. 'We can't abide the fruit in this house,' he explained, 'but the bush has been there long years and it's a shame for them to waste or the birds to have them, if your Missus can find a use for them.'

We said our thanks and set out for home. The flock was tired and not inclined to wander. Not that they had much chance. The collie darted around, worried if they dared bend their heads to snatch at the grass on the verges or if the lambs ran up the banks thick with cow parsley, willow herb, lady's purse and thin-stemmed, yellow buttercups.

A car came along and she held the sheep against a field gate as the vehicle inched past. At the mouth of the lane she whipped round and turned the flock towards home. The side hedges were sparse and broken in places but she allowed none to go through. The sheep knew where they were bound and quickened their pace.

In an adjoining field one of my neighbours was bargaining with a local butcher who wanted to buy his lambs in one job lot. (Lambs were often sold in this way rather than taking them to market.) My arrival gave them some relief from tossing about prices, offers and refusals.

'Been dipping?' he asked.

'Old Jonathon's place.'

The butcher joined in. 'How is the old crook? Still making that cider of his?'

I must have looked a little frayed round the edges. 'Marvellous stuff. He should put it on the market.'

They both laughed. 'If he did, none of us would walk straight,' the butcher said.

I left them and went after the flock. They were already grazing when I reached Egerton.

27 Cashing in the lambs 

Ours, I am afraid, was in some respects not a very conventional farm. For instance, take our two orphan lambs, Charles and Freda: they were a confused pair. The locals looked a bit hard at them and even harder at us, when they came visiting and noticed the lambs were wearing dog collars; or, perhaps, when they passed them being taken for walks by the kids and their friends, trotting along on leads very happily like well-bred French poodles.

On one occasion Ellis's son-in-law, Thomas, arrived to find Freda primped and prettied with pink ribbon bows decorating her woolly head. They had been put there by Vicky's chums. That was odd enough, but what amazed Thomas was Freda's willingness to cooperate. She enjoyed being tarted up.

'What kind of a lamb is that?' he asked.

'This,' Nicholas Paul told him scornfully, 'is a play lamb.'

'Why don't you dress him up?' Thomas asked, pointing at Charles.

There was even more scorn in the boy's voice. 'Because he's a boy lamb and boys don't wear ribbons in their hair.'

Thomas almost forgot he had come to borrow the tractor for a few hours.

The story got around. At the following Monday market, tall Stan Roberts – whom we only seemed to meet at sales – twitted me. 'They tell me you're putting dresses on the lambs now.'

'Not all of them, only the girls.'

He grinned, but two passing farmers almost dislocated their necks trying to get a good look at me. They obviously found it hard to believe their ears.

Naturally, when we took the flock to be dipped, Charles and Freda were exempted. They had no liking for the rough company of the flock. They lived in the garden, hanging around the kitchen door hoping to catch a glimpse of their 'Mummy' and following Shirley like pet dogs when she did appear. If she was late with their bottle, they butted the door, and if that brought no response, they stood on their hind legs to look through the window.

Several times, as they grew bigger and more independent, we took them down to the flock in the hope that they would elect to join. Not them. They were not even mildly interested. They followed us back or, if the gate was closed on them, they stood baaing in anguish as we left and then promptly looked for a way back to the house. Without fail they succeeded, sometimes by very roundabout ways, in finding their way home.

This being so, it was something of a shock to walk out one morning soon after the dipping and find the pair of them apparently mixing with the flock of their own free will. We had brought the sheep up to the field near the house to drench them against worms and other stomach parasites. (Sheep need to be dosed periodically if they are to get the utmost out of their feed. They get infected by larvae from worm eggs excreted by other sheep and picked up when grazing. 'A sheep's worst enemy – is another sheep,' Old Jonathon used to chant.)

The drenching operation came as a nasty surprise to our orphans. They joined in because they had no choice. The collie raced round the flock, rounded them up and put them in the collecting yard where we were going to work. Freda and Charles were swept along in the rush. I spotted them just once, frantically trying to move against the tide and escape. Freda stood out because the kids had wound pink ribbon and coloured sewing silks in her collar.

A drench gun which could shoot a measured dose of medicine down ten sheeps' throats before needing a refill helped speed our work. The routine was: a quick squirt, a look at their feet in case of footrot, and a dab of red marked fluid on their foreheads to show they had been treated.

As usual the sheep did nothing to make our jobs any easier. They dodged and backed and crammed together as tightly as they could. But gradually we worked our way through, thinning them out until only a bunch of lambs remained crushed together in the furthest corner.

One by one these were plucked out and dosed until finally, surprise, surprise, underneath their feet, being used as woolly doormats, we found Freda and Charles. They climbed painfully off the concrete and baa-ed their indignation at the outrageous way they had been used. Every movement and sound conveyed their self-pity. 'Well, well,' John said. 'fancy meeting you here.'

They took their medicine quietly, even gratefully, but afterwards declined to rejoin their rough companions. Instead they stayed at our heels. When the yard doors were opened and the ewes and lambs poured out into the field, they went without the orphans. The pair watched them go and then hobbled stiffly back to the house to await the emergence of their beloved 'Mummy'. For the moment at least they had again opted out of being sheep.

The sequel came the following morning when the cows came in to be milked. There, right in the middle of the herd, were the two lambs. It appeared they had elected to be cows rather than sheep. For their part the herd – Gaffer, Whitey and company – looked at them curiously but made no protest. They did not operate a closed shop. They believed in a policy of live and let live.

Unfortunately the tightness of our milking schedule allowed me no time to try and explain the essential difference between lambs and dairy cows. They were chased out under threat of my rubber boot. The last I saw of them that particular day they were limping disconsolately through the barley, in the middle of which they had made themselves a hideaway, no doubt to ponder the harshness of the farming world.

Our first profits from fat lambs came about three weeks after the dipping when we took the first batch to market. It was a big occasion for us. Since we were inexperienced in judging weights, we hung a spring balance from a protruding wall beam in the stockyard, wrapped likely lambs in a sack and hooked them up to see just how heavy they were. The target was lambs over eighty pounds with plenty of meat on their frames. There was little difficulty finding them at this first selection and we soon sorted out fourteen and loaded them into Old Lil.

Shirley waved us off and we left the farm high with hopes of good prices. The van's diesel phut-phutted cheerily and from our seats John and I had excellent views over the fields which stimulated discussion and criticism of our neighbours' stock and work. Even Old Lil's rattles seemed more tuneful than usual. On the main road friends in more well-to-do vehicles, some towing expensive stock trailers, overtook us with mocking honks and waves. But, never mind, we were going in the right direction and had every chance of getting there.

On arrival we joined a queue of vehicles bringing lambs for sale, and in due course we backed up to the unloading ramp. A cheery man in PVC leggings asked, 'How'd you want them drawn boss?'

'Two pens, seven each.'

He cast a quick glance over the lambs John was pushing out of the van. 'Can't do that, only thirteen here.'

'Fourteen, I counted them in myself.'

Another helper arrived and laughed. 'You need to go back to school, boss. There's only thirteen here.'

They were right. There were only thirteen. 'One pen of seven, one of six, then,' I said, completely baffled.

'Never mind,' the first man said. 'Even if you can't count, you've growed some good lambs.'

John was as bewildered as me. We watched the stockmen sort the lambs into two groups, run them through the weighing machine and pen them. I filled in the form necessary to claim any subsidy that might be forthcoming and we went to grab a quick cup of tea before the sale started. Neither of us mentioned the puzzle until we had swallowed a couple of mouthfuls.

'The van door was shut properly,' John ventured at last. 'There's no way they could have opened it. If they had, they'd have all gone, not just one. We must have miscounted.'

Tall Stan came in and boomed across the room. 'First ribbons for lambs. Now you've got so many you can't count them, I hear.' The tale of the missing lamb had leaked out. There was nothing to do but grin and bear it.

Our two pens sold well. We collected £10 cash and a cheque for £79.60 from the paying out office and went home. The money delighted Shirley.

'Don't worry about it,' she said when we described the mis-count mystery. 'It must have been all the rush and worry.' She spoke in that special tone women reserve for very young children, the mentally retarded, or elderly folk showing signs of senility.

But all was not yet finished. There was a final act to the riddle. It was played out that evening when I walked round the dairy cows after they had been milked and turned out to graze. The collie pup, which had been following at my heels, suddenly took off towards the lane gate. Easy to see why: a fat lamb was trying frantically to get through the bars and on to Egerton land.

No doubt about it, this was our missing lamb. He had our identifying mark and once through the gate he made for the bottom field where the flock was spending the night. His target was an Old Clun Forest ewe which was as pleased to see him as he was to see her.

So we *had* loaded fourteen! But what had happened? One possibility was that the van had hit a bump somewhere en route to market hard enough to jar open the rear doors so that the lamb had fallen out. The doors could, just conceivably, have swung back and refastened. Stranger things happened with Old Lil. If this was the explanation, however, where had the lamb spent the intervening twelve hours? How far had he come and how had he found his way home? No one ever provided any answers to these questions.

When Tall Stan heard the story, he said, 'You'm so lucky, if you fell in the river, you'd come up with pockets full of fish.'

At least the lamb's adventure earned him a reprieve from the market, even if it was only short. Lamb prices stayed good and as ours reached saleable weights, we took them in. For quite a few weeks our old van was in the queue for the unloading ramps as we added our little bit to the nation's food production.

28 The calves go missing

The drama of the missing calves came just at a moment when our heads were stuffed with hay. It was dreadful timing. You could, Old Jonathon said, almost hear the grass growing. Before long it would be crying out to be cut. Any day now would bring the first rattle of the mowers at work in the fields. In short, the hay harvest was upon us. The locals went about with fingers crossed, waiting for the right run of weather to trigger off operations. We crossed just about everything possible and walked round hoping we would be able to cope.

On the advice of friends we had bought a secondhand mower for £60 and, for another £49, a weird but efficient machine called an Acrobat which turned the hay for drying and collected it into rows for baling. They were standing ready and waiting to be used. As usual, we read everything available and were shameless in picking the locals' minds. The crucial decision was when to begin cutting. Once a start had been made there was no turning back.

'When you knock it down, you've got to get it finished and inside as quick as you can,' Griff said. 'Unless you wants to leave it in the fields to rot.'

So with this much in my thoughts, there was some excuse for my being more than usually obtuse when, as the last cow undulated out of the milking parlour and I set the cooler chugging round in the churn, Shirley burst in looking like a one-woman Greek chorus.

'Our calves: they've gone. All of them!'

This was the stuff coronaries are made of. 'Gone? What do you mean, gone?'

She threw up her hands theatrically. 'Just that. We've been robbed!'

By this time we were trotting down the stockyard and it was, incredibly, true. Where there was usually a great moo-ing and bawling, there was silence. Nothing.

We had put the sixteen oldest calves in the big open-fronted pen at the bottom of the yard to give them more space. They had certainly been there the night before when I walked round the buildings before turning in, but this morning, when Shirley went down to feed them, there was no sign. The front fence was flat on the ground. The pen was empty.

Not one calf remained. Taffy and Ferdinand, Alice Capone, Beauty, Black-Eye, Mother's Pet, the whole mob, had vanished. Even the youngest bottle calves in the closed pens were silent until we rushed round in a panic, flinging open doors to check. Then they stood up and bawled to be fed, but today they would have to wait.

John arrived. He had phoned the nearest neighbours. 'Nobody's seen anything,' he said, 'nor heard anything. Not a sign.'

Time was passing. The milk had to be lugged to the top of the lane. The lorry was waiting when I arrived.

'Nothing on the roads I've been along,' Jock the driver said. There was talk of two black heifers on a road towards Whitecliff but that's seven or eight miles from here and they don't sound like yours. Don't you worry, though, they'll be around here somewhere. If I hear anything on the round, I'll get someone to give you a call.'

When I got back to the house the other two were trudging back from a search round the fields. Not a trace. Our precious calves had gone. But where? Sixteen calves don't just take off without a trace. It needed thinking about. Breakfast was waiting but no one could eat.

John and I climbed into the 1800 and drove along the local lanes. Round and round they twisted. Branches went off in all directions. It was a tarmac maze. Folk we spoke to were no help. No one had seen anything. We began to think in terms of theft. There had been several stories in newspapers recently about stock being rustled, butchered and sold to restaurants and hotels. A farmer at nearby Sollars was said to have lost six hefty bullocks only a week before they were due to go to market. Perhaps our sixteen juveniles had gone the same way. I was despondent enough to believe the worst.

The lane gate had been open. It quite often was. A lorry could have free-wheeled down the lane but it would still have had to be driven back up and someone must surely have heard something. Noisy old Moses the gander and his biddy, Martha, would have had something to screech about. But supposing the calves had been quietly walked up the lane to the waiting lorry? That was more feasible.

I rang the local police station and began, apologetically, to tell my tale of woe and our suspicions.

'Look,' I said, 'this sounds a bit far-fetched but we've lost sixteen calves. Last night. It's just that we can't see how they could have gone by themselves and no one seems to have seen anything. Then there's been all this talk about rustling . . .'

A little ridicule from the Bobbies would have been welcomed at this stage. Instead, to my horror, they were quite prepared to take the idea seriously. 'There's a bit going on,' the sergeant said. 'We've had a circular round telling us to keep our eyes open for anything strange. I know your place. It's a bit isolated. It could happen. I'd better come along and take a look round. In the meantime, if you hear anything, let us know and we'll get back to you if we hear anything.'

The exchange more or less confirmed our worst fears. Shirley was stunned at the thought of her 'girlies' coming to such an end. Bad enough to know that they were doomed eventually to become beef steaks but they were only babies now. For myself, the financial implications of such a theft were horrifying.

John, as ever, reserved judgement and preferred to wait for developments. But the two kids were fascinated by the idea of rustlers. Nicholas Paul visualized them as masked, gun-slinging riders with the regulation big black hats. He was expanding the idea when Shirley realized they had missed the school bus. She went off to take them in the 1800.

Finally, late in the morning, I rang Griff at the Forge.

'No, heard nothing,' he said. 'But don't you go worrying now. Likely they'll turn up . . .'

An hour later he came back. 'They'm at the Ford Farm. The postman saw them there. You'd better give old Johnnie Burton a ring . . .'

The farm was about three miles from us. I rang the farmer.

'I don't know how the hell they got here,' he said. 'They were in the field this morning with the gate closed on them. I thought Father Christmas had brought them. Better come and get them before they'm off again.'

It was surprising how the day suddenly brightened. The three of us raced for the car.

As soon as the runaways saw Shirley they began to bawl. Taffy and Ferdinand came up to be fussed, only to be pushed aside by the bigger calves. They were all there and all safe and sound.

'Someone must have turned them in off the main road,' Johnnie Burton said. 'They must have been on the top road. My neighbour Reggie said they had come down the lane, judging by the amount of muck on the ground.'

The 'lane' was part of an old green road which streaked across the local countryside. It sounded as if the sixteen had moved in a huge circle, stopping now and again to graze on the verges. We were lucky to get them back undamaged.

They were tired. The little ones drooped on the journey home but, with John leading, me behind them and Shirley in the car bringing up the rear, they had to keep going. As we passed the Forge, Griff appeared, cloth in hand, and asked, 'Everything all right, then ?'

'No harm done,' I said. 'We thought they'd been rustled.'

'Like in the pictures ?' he asked intrigued.

'Well, not exactly, in a lorry.'

Matthew came out behind him. 'Jacky thought they'd been rustled,' Griff said.

'I heard him,' Matthew said. 'Not much fear of that happening. Not now. The butchers 'ull wait until you gets them to market, then steal them.' From which I presumed that beef prices were down.

It was a scorching hot day. Before we reached home everyone, calves and people, were feeling the strain. I had begun to think in terms of long cooling drinks when a car came up behind us in the lane. It was Andrews, the 'cutter', who was going to castrate the bull calves and dehorn all of them for us. There was no question of putting him off. He was an extremely busy man with a constantly full order book.

'Be with you in about an hour boss,' he said. 'OK?' He called everyone 'boss'.

Things, I assured him, would be ready.

We turned the calves into the big covered yard and they made a bee line for the water tank. John and I rushed about man-handling hurdles to section off a corner of the yard to confine the calves for Andrews's attentions. Getting the calves into the pen was no difficulty. They had little spirit left. I knew how they felt. While I dragged weary legs towards the house and coffee, John, bouncy with the resilience of youth, took a small axe and went to cut kindling for Andrews's fire tins.

It seemed only minutes after I had sat down that the telephone rang and Willem said, 'Andrews has finished here and is on his way to you. You've got the calves back, I gather.'

'Down at Johnnie Burton's. No damage done.'

'You'd a bit of luck there, then,' he said. 'There was a chap at Neenton had a growed cow's leg took right off with a lorry one night. Bloody daft drivers!'

I remembered one night on a motorway near London when I nearly slaughtered myself and a straying pony. But that was in my city time. 'Bloody daft drivers,' I agreed.

I rang the police to let them know all was well. They sounded a trifle disappointed at the anticlimax but rang back later in the day to say they had received a report of the calves being seen near a village miles from where we had collected them. From our pooled information, the helpful sergeant and I concluded that our runaways had covered between eight and ten miles in their wanderings before ending up in Burton's field.

I was still shaking when Andrews arrived. He was a broad, muscular, middle-aged man with heavily developed forearms and big hands. There was an air of competence and calm about him.

'Been having a bit of trouble, boss?' he asked.

'Damn calves took off.'

'Never you mind, you've got them back, boss. That's all that matters.'

He was busy setting out his instruments in a dish containing an antiseptic solution. The cauterizing irons were heated in the fire tins, fuelled by the wood John had chopped. His voice was quiet, so as not to disturb the calves.

'Come along, my beauties,' he crooned, and before the chosen calf could escape a rope halter had been dropped neatly over its head and tightened. The next step was to secure it to a heavy corner post for his attentions.

The calves, particularly the big ones, might be tired and footsore, but they had no intention of offering themselves easily. They milled around, crushing into the furthest corners of the pen, throwing their heads all ways to dodge the halter. One after another they had to be pushed and pulled and manhandled into position. One of the biggest trod on my foot. I was wearing Wellingtons and the result was two lost toenails.

By the halfway stage I was awash with sweat. There was a taste of salt in my mouth. My legs were bending and my arms seemed about to drop off.

'They can get a bit frisky,' Andrews conceded, seeing my state. 'But never you mind, boss, you've got a nice bunch of calves here. They'll do you right.'

When it was finished we went into the house for coffee and to settle our account. He had a small farm of his own with a Jersey herd which his wife managed, but they were thinking of going over to the bigger-yielding Friesians. It was interesting talking to him because he had a great fund of stories. But it was a mistake: I should have stripped off and towelled myself dry. By the time he said his farewells and went off in his car, it was too late.

The next morning I paid the penalty. The alarm exploded at 5.30 a.m. and brought me, reluctantly, from under the bedclothes. I sat on the edge of the bed and reached for a sock. Everything promptly blacked out and the next point of clarity was an anxious Shirley bending over me as I lay in bed. John had done the milking and the doctor was coming to see me. I had a temperature of 103 and possibly pneumonia. What a sense of timing! Just when the hay needed to be cut.

When the doctor arrived it proved to be a female of the species and a very tough one at that. She injected me painfully with some vaccine or antibiotic and said, 'There. Now you know what animals feel when you inject them.' Perhaps she didn't like farmers. I preferred to think that it was nothing personal but I could have been mistaken.

After making it plain that she considered the whole affair my fault, unnecessary and a waste of her valuable time, she informed me in clearly defined terms that I was to stay in bed at least a week. She drilled Shirley in what needed to be done, added a few final threats and departed. I felt trapped, depressed and very sorry for myself. A week in bed at this time of year was out of the question. The weather was not likely to continue indefinitely in our favour.

Her attitude had affected Shirley. After her car had roared away, I got out of bed to reach the toilet which was along the landing. It must have been the effects of the injection. I came to being dragged along like a limp sack by Shirley and John, having, apparently, passed out again. When I was back in bed, my wife recited what she could remember of the doctor's instructions and ended by shaking her fist and threatening to murder me if I stirred a foot without her permission. She was very annoyed.

John's schoolboy shoulders had to take the load. He did so without a moment's hesitation. He had just finished O-levels and his school had agreed that he need not finish out the term. On the day of the medical visitation the BBC weatherman forecast a prolonged period of hot, dry, sunny weather and John, and just about everyone else, started cutting.

Fortunately he was a very proficient tractor driver. When we had acquired the David Brown I had simply shown him the controls, how to stop and start, and left him to it. He had spent hours driving round the top fields and up and down the lane. Now it paid dividends.

Lying abed I could hear him working. It was exasperating. My being out of action meant the mowing had to be fitted in with other jobs and it all added up to an extremely long, tiring day. The tiredness showed in his face, his fair hair was plastered to his forehead by sweat, but he stuck to the task without a word of complaint or a request for help.

Then our luck took a turn for the better. Some London friends arrived, unexpectedly, for the weekend. They were suburbanites born and bred. For them the countryside was simply a vast picnic site. Now they astonished us by pitching in to help with a gusto I would never have suspected. While hubby took his turn on the tractor and proved very deft indeed, the tall, elegant wife helped with the feeding and cleaning, pigs and all.

'They,' she said, referring to Dorrie and Dorfie, 'are preferable to a lot of my dearest friends.'

Their kids were left to fend for themselves and, probably as a result, had a whale of a time. The younger boy spent most of the weekend in the henhouse waiting for the hens to lay. At periods throughout the day he appeared in the kitchen and presented newly laid eggs to Shirley. He was so excited by it all, words would hardly come. His bigger brother spent his holiday comforting a rather quietened Ferdinand who was getting over the cutter's visit but still welcomed a little extra fussing.

It did nothing for my ego to discover that my absence as a host was hardly noticed. They even managed to drive into town for dinner and returned, Shirley as well, full of high spirits and not even prepared to stop talking when I tried to explain how I felt. They departed on the Sunday evening vowing that this country life was for them, the air was cleaner, the food tastier, they would dispose of their suburban detached and invest in a farm. Shirley undertook to let them know if I survived.

By the fourth day I was bored and well enough to shake off my self-pity, override Shirley's protests and quit bed to lend a hand round the place. It was startling to discover just how weak I was, but work helped and in any case John continued to bear the heavier load. It did not, though, endear me to our medical martinet when she arrived. She left me in no doubt whatsoever about her opinion and did not even try to understand my concern for the hay.

 29 While the sun shines 

Everyone, it seemed, was making hay while the sun shone. The entire district was alive with tractors and turners, trailers and balers. The Forge was almost deserted. No one had time for anything but the grass harvest.

Modern haymaking, more's the pity, is very much a mechanized process and far removed from the popular concept of cider-drinking rustics, horse-drawn wains and rosy-cheeked wenches dispensing goodies from chintz-covered baskets. But some things are unchanged. The smell of the new-mown hay remains, the sun is still hot on a man's face, and the hum of insects busy on the clover is the same. There is something basic, something special, about harvesting the grass. It is the farmer's gamble against the weather and when it comes off the finished product is the gift of the sun.

This year the sun was on our side. Day after day it appeared on schedule. From Egerton we could see a dozen machines at work. Some were so far off that the sound of their engines did not reach us and they looked like red, crawling insects. But they were machines all right, and steadily, remorselessly, they edged round the fields and the mower blades scythed through the tall-stemmed grass and felled it in rich, regular swathes.

We cut and then, as the grass dried, we changed mower for turner and sped round the fields to flip the drying swathes over and thus offer all sides of the grass to the sun.

The quicker the making, the better the hay, the locals said. June hay they judged to be best of all if it could be made without being rained on. Well, the month was June, and this year everyone, even we, could not help but make good hay.

Towards the end of the week, the quietly spoken Price from the top of the lane arrived to check if our hay was ready to bale. He picked up a handful and crushed it, held it under his beaked nose and sniffed, bit it and, finally, screwed it into a knot. When it passed all his tests, he nodded and said simply, 'This field's ready. The others can do with another couple of days. I'll come here tomorrow.'

He looked tired. Like us he had his ordinary working routine to get through and his own hay to make. In addition he contracted to bale for people like us who did not possess the necessary machinery or for bigger farmers who might be able to afford the machine but needed the extra labour. Like shearing for the young men, haymaking gave him a chance to make money but not one penny came as a gift.

Next day, mid-morning, when the sun had had time to dry out any night damp, he appeared driving his big Fergusson tractor, towing the baler and with a bale sledge bumping along behind. John was already at work in the chosen field, having set the turner's pronged operating wheels so that they gathered the hay from its spread-out drying position and left it in rows ready for baling.

Price was not a talking man. Minutes after arriving he was following in John's wake. The low, squat baler gobbled up two rows at a time, compressed the hay into bales, tied them with binder twine, and ejected the finished articles to be scooped up by the following bale sledge.

The sledge was descriptively named. It was very like the dog sledges which appear in films involving huskies, fur-hatted trappers and Mounties who shout 'Mush' at frequent intervals. But this sledge was made of tubular steel. It slid over the ground gathering bales until there were seven, whereupon Price tugged a control cord, released a swing gate at the rear of the sledge and left them standing on the hay stubble. The advantage, and it was a considerable one, was that the bales were thus gathered ready to be stacked for lugging.

Shirley, having finished her own work, had come down to help me stack. It was much easier working as a pair. A bale of finished hay weighs between 40 and 50 lb. Newly made hay still has a lot of moisture and weighs more. It became a family joke that the first time Shirley tried to lift a bale she ended flat on her face without moving it.

Apart from the weight there is also the binder twine which ties up the grass package. It cuts very painfully into your palms.

My hands had hardened, but I was still grateful for a pair of old gardening gloves. Some of the locals grabbed a handful of hay to act as a pad. Shirley was wearing rather snazzy kid driving gloves.

We worked along the rows. A bird's nest, presumably a meadow lark, had been harvested with the hay and tied into one bale. There was a patch of feathers.

'If you had told me I could help stack hay, we would have tried farming sooner,' Shirley said sarcastically.

'It doesn't do to shout these things about,' I explained. 'There'd be a queue of women if it was known.'

'Like hell there would,' she said in a very unladylike way.

In fact, although it was hard on the back and hands, it was not unpleasant work in the sunshine. The trouble was that both of us had a tendency to indulge in morbid calculation. Thus the meter on Price's baler showed there were 1,088 bales in the field. That meant, at seven bales a stack, about 155 stacks. We had done twenty: another 135 to go . . . Better really to close one's mind to figures and work mechanically – bending, lifting, stacking, moving on.

Once baled and stacked the hay was less vulnerable to the elements. It could be left standing in the fields to go on 'making' while the farmer got on with the next patch. The curing process would continue in the barns. Indeed the heat generated in the barns was quite astonishing, although modern baling practice has reduced the risk of spontaneous combustion. Even so it was a danger to be considered. During the following winter I opened several bales and found them hollow, eaten out as a result of combustion which, fortunately, had been limited to the particular bale.

Altogether that summer we got 2,565 bales: 793 off the eight-acre and 684 from the House field, in addition to the bottom field's product. 'That amounts to 366 stacks,' the calculating Shirley announced. 'Or, roughly, taking a bale as 40 pounds, about 46 tons.'

Each of the three bays in the barn could take about 800 bales when empty but some hay remained from the previous summer. My inclination was to build a stack outside and cover it with a ricksheet but the locals advised us to try and get it under permanent cover. We listened to them and, to the utter delight of the cows used one section of the big covered yard where they were going to winter. The hay would be dry and would also act as a windbreak, we thought.

Just what accomplished thieves cows are, we soon discovered when we brought them inside at the end of the autumn. My first attempt at protection was a wire-netting screen. It seemed reasonable to suppose they would not be able to get their tongues through the holes to reach the hay. What a joke that proved to be! Not only could they get their tongues through, they ripped holes in the wire and tugged out whole bales. How they managed without cutting their tongues, I do not know. But manage they did. We substituted tin sheeting and that did stop them, although they never gave up trying to find a way through.

But that came some months later. The job now was to get the hay inside.

Back and forth we trundled between field and barn, riding when the trailer was empty, walking behind when it was loaded. The little two-wheeler trailer I had acquired at a sale for £8 could hold, say, 100 bales. We did manage on one occasion, with an enthusiastic friend, to build a towering 129-bale load, but it had to be secured with ropes and even then the trailer overturned and spilled the lot.

For John and me, working by ourselves, the average load was about 90 bales. This, when we saw the prodigious carrying capacity of some of our neighbours' trailers, seemed very paltry. It might have been advisable not to depress ourselves by working out how many times 90 would go into 2,565. But we did: 28 trailer loads.

We did not possess an escalator so the bays had to be filled by building a series of steps and planes. It would be a conservative estimate to say that every bale was handled at least three times between the start and finish of the harvest.

However, the weather held and held, and gradually the fields emptied. The bays filled up. The yard stack was built. Finally there was a marvellous moment when we brought in the last load. A V-shaped space had been left for it in the front bay and now we filled it, bale by bale, until John dropped the last one into place, stamped it tight and climbed down the ladder to join me.

'Somebody ought to make a speech,' he said.

That was how I felt but there was no one else about.

An hour or so later Howard arrived to see how we were coping and marched, looking like the truculent ex-sergeant he was, to inspect the filled barn. All three of us walked down to look at the empty hay fields. The cows were already gleaning the bits of hay left behind. Like new grass it had disastrous effects on their insides; it could also 'blow them up' if they were allowed to over-indulge. None of this lessened their appetite for it.

'That looks a lot neater than it did a day or two ago,' Howard said. 'It's been a most wonderful stretch of weather. Never known a summer quite like it. Not since I was a boy, anyhow.' Summers are brighter and last longer when we are boys.

From there we all went up to the Forge in his car to get celebratory beers and a cider for John. Griff was polishing his glasses. 'Got it all in, then?' he asked.

'Just brought in the last bale,' I said, trying hard to be nonchalant. 'How's others doing?'

'Most have finished. One or two were a bit slow off the mark but most have finished. There's a bit of good hay about this year and a lot of it.'

'Better too much than not enough,' one of the Forge regulars said, 'but you can offer your life that come next March we shall all be reckoning how much we've got left and how long it's got to last.' He offered a drink and got the round and we sat discussing haymaking – he from the experience of sixty-odd years, John and I from the lessons of one season.

The weather broke at the end of the week. The spatter of rain against the bedroom window brought me instantly awake and sitting up.

'What's the matter?' Shirley asked, wakened by my movement.

'Listen. It's raining. Hard.'

She sighed in exasperation. 'It doesn't matter now, does it?'

It didn't. Not for me, anyhow.

 30 Midwife to a sow 

Dorfie and Dorrie, our two amiable four-legged, pork-production factories, continued to thrive. The success of their visit to the boar became increasingly apparent as the summer progressed. Their stomachs seemed to hang a little lower every day but, pregnant or not, they still foraged, chomping all things edible, no matter what, pushing their long snouts into anything and everything.

Occasionally they quarrelled and shrieked at one another with shrill, porcine vituperation, but mostly they got on very well and moved round in tandem, Dorfie leading, Dorrie in the rear. When our paths crossed theirs, we exchanged greetings and grunts and continued our respective businesses. Sometimes, when we had time, we would scratch their backs. A piece of rough board made a very effective instrument.

Eventually, though, just moving about began to get too tiring and they rested a lot. An old dunghill, the muck long weathered into black loam, was a favoured spot. Here they lay, basking in the sunshine, their enormous bellies spread out like fleshy tablecloths over the rough surface.

Everyone seemed to be interested in their progress. Pigs might not be considered a sound agricultural proposition at that moment, but they hold a special spot in farming affections.

'How's them sows getting on?' Howard would ask. 'You need to keep a close look on them or you'll have trouble.'

We needed no prompting. As their production dates began to loom up, we kept testing their nipples to see if there was any sign of milk.

'Within twelve hours, that's when they'll come if you finds milk,' Aaron 'from up the mountain' told us. 'They might be here quicker, but they won't be longer.'

Three days before the due date we put Dorrie into the farrowing crate. She waddled in after a bucket and a few pignuts and hardly noticed the bars. Some of the locals thought we had moved her in too early, some thought she should have gone there before.

The crate was a contraption built of tubular steel and set firmly in the concrete floor. Once in, a sow could not turn round or, more important, flop down. It was the 'flopping' that farmers feared. There were a dozen cautionary tales about sows that had smothered entire litters.

There was an echo of 'hard times' in the story Aaron told. 'I was just a bit of a nipper and we was so hard up, every penny was worth a pound,' he said. 'There was the rent to pay and the old man counted on settling it by selling this pig with her litter when it came. But, like it usually happens, there was only five pigs where we could have done with ten. My father was biting on that disappointment when the old girl had a stroke or heart attack or something and falls flat on the whole bloody lot. She just lay there and we couldn't get her up or get them from underneath. They was all goners . . .'

'What did you father do ?' I asked.

'Ah, that poor man,' Aaron said. 'He was no saint but he had a fear of the Almighty and was worth a bit better treatment than he got. He went right down to his knees there in the sty, I thought he'd gone the same as the old pig, but he prayed. "God" he said, "let her live . . . at least until the butcher comes, because now we canna sell her, we'm going to have to eat her."'

Someone handed him a pint and asked, 'Did her make it ?'

Aaron was silent for a few seconds, busy with the memory. 'Oh, her come through,' he said at last. 'She lay there close to death croaking and groaning for an hour or more and then she got back to her feet. The next Wednesday the butcher, old Watkins, the present one's father, come and done her. I remember it well; no freezers then, chitterlings and spareribs, my old lady salted a lot of the meat. We lived better that winter than we did one or two others. It sounds daft, but I loved that old pig.'

Everyone was quiet with their own thoughts until I asked, 'What about the rent ?'

One of Griff's many relatives laughed. 'Same as the rest of us, I suppose. There was no money about anywhere. We was on the estate then, the little White Farm. It was the old Lord then and he knew what was going on round him; he didn't mind waiting a bit when it was right.'

Perhaps Dorfie was lonely on her own. The next morning she came quietly into the milking parlour and nudged me with her snout. So while John looked after the cows, I led her into a crate with a bucket. No problems. She was even greedier than her sister. Both of them lay splayed out happily on the concrete floor, occasionally grunting to each other, to wait their happy events.

Egerton was well equipped for pigs on a small scale. The farrowing unit held six crates, each separated from its neighbours by boards which could be lifted to allow piglets to move between sows for multi-suckling. There was water and a feed trough at the end of each crate. A creep area ran the length of each crate so that newborn pigs could retreat there after feeding and sleep under the suspended infra-red lamps, warm and safe from the danger of being rolled on by their mother.

For the kids it was a magic time. They sat on the edge of the crates and stroked the great, bumpy-with-piglets bellies. Dorfie found this particularly soothing and encouraged them with short, happy grunts.

Nicholas Paul sat down for dinner one evening and announced, 'Mine's stuffed full of pigs, there must be at least twenty.' The game, apparently, was to try and guess from the bumps how many piglets there were inside each sow.

It was on the third evening she had spent in the crates that John tested Dorrie's teats and, sure enough, as per the book, out bubbled a tiny white pearl of milk. 'They'll come some time tonight,' he said, almost as awestruck as me at the implications of the glistening drop.

We decided to adopt a rota system to check on the sow during the night. Shirley was co-opted to serve. There might not be much we could do but if anything did go wrong, at least someone would be aware of it.

As luck would have it, the first piglet was born about ten o'clock, just as we were thinking of going to bed. John had gone to check. The sow lay quiet, breathing heavily until, suddenly, she convulsed and right before his eyes the first piglet appeared, in the birth sac, umbilical cord still attached. It broke through the transparent sac, the cord came free as the new arrival struggled and pulled, and with some three inches of it dragging beneath, the piglet began to search for the sow's teats. As with all young animals, a piglet's first feed is vitally important for survival and growth. The umbilical cord dried and shrivelled and ultimately fell off.

The sounds that came from the newcomer were more like a duck's quacking than anything else. There was a quick response from the sow. She gave a series of short, encouraging grunts which must have helped guide the firstborn. He, for it was a hog pig, hunted towards the sound, climbed over her bottom back leg and, just minutes after birth, was sucking at a teat, eyes closed in ecstasy. The sensation must have been pleasant to the mother because she lay very still, grunting contentedly.

Having seen the first one settled and feeding, John raced up to announce, 'They have started to come.'

'The pigs?'

'Well, they might grow up to be something else, but they look like pigs at the moment.'

His wit was wasted. All thoughts of bed were abandoned. Next morning and milking seemed light years away. All three of us hurried to witness the farrowing and watch as five more arrived in very short time and stumbled about after a teat.

As they came, Shirley picked them up, dried them on an old towel and placed them on the teats.

It was gone midnight before the last piglet was born. He was tangled in some of the afterbirth and must have suffocated had we, or rather John, not been there to get him free and set him en route for the milk bar. The final total was fifteen, with the last one, surprisingly, easily the biggest.

In size the litter ranged from the last, a whopper, to the runt, a tiny thing which kept being pushed off its feed by more robust brothers and sisters. 'I wouldn't give much for his chances,' Shirley said worriedly, but there was little that could be done for him, and at least the other fourteen were all healthy, vigorous creatures.

Sow's milk is rich, heavy stuff. By the time the last one arrived, the earliest arrivals had taken their fill and found their way under the infra-red lamp. We had layered the creep with packed-down straw and they lay tumbled together under the circular light in a pink, contented heap.

We fed and watered the sow and then went back to the house. It was like being kids all over again and getting a special, unexpected gift. In the living room we sat in front of the fire drinking coffee until Shirley noticed the time – ten minutes past two – and we all realized we were bone tired and went up to bed. A little of the elation was lost a few hours later when we found the runt dead under a heap of piglets.

Forty-eight hours after her sister, Dorfie produced. Naturally, being her, she chose to begin the operation in the early hours of the morning. The first three arrived one after another like peas from a shelling machine, but then there was a hang-up. The fourth refused to join us. Something was obviously wrong. The sow grunted and strained but nothing happened.

'You will have to do something,' Shirley said helpfully.

'Such as?'

'Midwifery or some such thing.'

It was one of those many moments when it would have been nice to have someone with a little experience around the place. But there was no one. So, having glanced quickly at the chapter in the book and found it useless, I washed my hand, lubricated it with soap, and tried feeling inside the sow for the reluctant one. I touched something, presumably number four, and presumably dislodged whatever was stuck because, a few minutes later, he was duly expelled and started on his journey to the teat bank.

The final score was ten, against Dorrie's fourteen surviving piglets, but they were bigger and heavier which compensated for the difference in numbers. Many of our friends, Howard for one, considered ten healthy pigs to be about the ideal size of litter for a sow to rear.

Dorfie's performance was altogether noisier and more extrovert than her sister's. She huffed and puffed and managed to spread the whole affair over nearly four hours. By the time it was all finished, there was not much point in going to bed. Instead we had a cup of tea, did the milking and feeding and then went upstairs to grab a few hours' sleep.

Sow's milk is deficient in iron and, unless something is done, there is a danger that the piglets will develop anaemia. So, five days after their birth, we injected both litters with an iron compound. It was a quick injection into the rump, but naturally the piglets protested. This brought an immediate response from Dorrie which made us glad she was securely in the crate. Not so Dorfie. She was uninterested in the whole affair when it was her litter's turn.

It was intriguing to see the difference in the two sows. Dorrie was the perfect mother. Her piglets ran everywhere. When she stood up they jumped to try and reach the teats. She was terrified of treading on them and lying down was a slow-motion exercise.

Not so our Dorfie. She trod around as clumsily as ever and flopped down regardless of what, if anything, was underneath. There were some hair-raising escapes but, fortunately, her bulk ensured that the lower bars of the crate took the first impact and delayed her arrival on the floor long enough to let the litter get away. It was nerve-wracking to watch her, but the piglets learned quickly and were a lot nippier round the pen than Dorrie's offspring.

There was no need to ask where the kids were these days. They spent all their spare time watching the piglets, a spectacle rather like a mini-gladiatorial contest. The new arrivals wrestled and pushed one another about in trials of strength until they were too tired to continue and fell asleep under the lamp.

On several occasions worried mothers rang Shirley trying to locate sons who had been smuggled down the lane to witness these wondrous goings-on. John suspected our pair of charging admission. They certainly seemed well supplied with sweets.

Last job at night for us was to check that everything was well. Usually the sows were sleeping, but if not they would scramble up hopefully, looking for a late-night nosh. They didn't get it. They were already receiving more than fourteen pounds weight of pignuts daily.

We were warned of the dangers of keeping them too long in the crates. They could 'go off their legs' and lose the use of them through inactivity. So, five days after the last littering we moved both sows and their families to the follow-on pens. These consisted of an outer pen with feeding trough and water supply and an inner sleeping section. In addition there was a creep area, made inaccessible to the sows by bars, where the piglets could enter and get the special food pellets which supplemented the sows' milk.

One difficulty was that these pens were some distance from the farrowing unit. We had prepared everything possible in advance ready for the transfer. The sleeping quarters were strawed and this bedding was well tamped down. Sows are inveterate ratters. If they see something moving about under loose straw they're apt to snatch first and look at what they have got afterwards: a rodent is a welcome addition to their menu, but a pig has to be quick to catch them. We lost one piglet from a later litter in this way.

Shirley produced a cardboard box and we loaded the first ten protesting, screeching piglets into it and carried them away. Then, when we let Dorrie out she trotted alongside me, never taking her eyes off the box containing the last four, an almost human expression of anxiety on her snouted face.

There was something touching about the scene when she joined her litter in the pen with full access to them for the first time. The babies greeted her with a chorus of grunts and squeaks and milled around her feet. She towered above them and bent down to sniff them, presumably to check that no harm had been done. Then, this quick inspection completed, she called them to her and led the way into the sleeping section where they would be safe from our prying eyes. Half an hour later we crept in to look and found them all asleep. The sow was lying on her side with the piglets, all fourteen of them, cuddled against her for warmth and comfort.

Dorfie's exodus was altogether different. When we collected her piglets, her ladyship raised her head enough to see what all the commotion was about. Satisfied it was only us, she settled back again. As far as she was concerned we were welcome to the whole pestering litter.

Nor was she in any hurry to join them in their new pen. It took the usual bucket and a few food nuggets to lead her there and even then, when the door was opened and the piglets rushed forward expectantly, it looked for one long moment as if she were about to turn round and make a run for it. But the sight of the nuts we had thrown into the pen was too much: she went in, gobbled the food and took a longer, second look at the piglets, decided they were not so bad after all and led them into the sleeping section.

Both sows proved excellent mothers but Dorrie's mothering instinct was much stronger. When the piglets were big enough to be allowed out to forage with the sows, it usually ended with a harassed but happy Dorrie surrounded by two dozen demanding, grumbling youngsters while Dorfie wandered off on her own business.

The first day they did go out into the fields, we were worried in case the two litters got mixed.

'This should do the trick,' Shirley announced and produced a lipstick guaranteed to be 'kiss-proof'. We used it to mark an 'X' on Dorfie's litter's bottoms.

Unfortunately, while it may well have been kiss-proof, it certainly was not bottom-proof. By the time they returned to the pens, all the piglets had rubbed against one another so much, all their posteriors were lipsticked.

There was no chance of identifying which was which, so we simply put twelve in each sow's pen. Dorfie never objected. It's doubtful if she even noticed there were two extra for supper. She had plenty of milk and all were welcome – just as long as they did not expect her to play the hostess or the doting mother.

Just at this moment, with the farm apparently flooded with pigs, we felt the world was floating at the end of our particular piece of string. We needed someone to impress. No point telling the locals, they'd seen it all before, many times, and on a much bigger scale. Happily, in the midst of all this productivity, more visitors from outer suburbia arrived. For them, like us, a piglet was still something of a mini-miracle.

 31 Quiet day in the country 

Our friends – William, Maggie and their two children – were as welcome as Christmas crackers. Exactly what we needed: someone more ignorant than us about farming matters. In short, someone we could impress and someone prepared to believe our tallest stories.

They looked remarkably sleek and polished, replete with an urban affluence which afterwards prompted Nicholas Paul to say, 'Once, when we lived in the city, we were rich, but now we live on a farm and we're poor.'

Almost before their car engine had cooled, we whisked them off to see the new arrivals. They were satisfyingly awed by the fact that we, with only the help of two sows and Howard's boar, had been able to produce twenty-four quick and sturdy piglets. Even better, since they were deep-freeze enthusiasts, Maggie could also evaluate the potential of said piglets in terms of chops, bacon, legs of pork and other joints.

They walked the bounds of the farm admiring calves and cows, lambs and ewes. Seventy-five acres might not be a very big farm, but to our visitors, conditioned to the confinements of suburbia, it was an enormous estate.

When we got ready to milk that evening, Maggie asked, 'Would we be in the way if we came and watched?'

They were a wonderful audience because it was something quite outside their experience. Every movement of a cow's foot sent a shiver of apprehension through them and increased their appreciation of men daring enough to come within kicking distance. 'Perhaps,' my son suggested slyly, 'we should charge for admission or make a collection.'

It was what we could do as an encore that was worrying me.

They helped with feeding the calves and in the evening we sat round a log fire in the end lounge, drinking Shirley's potato wine and discussing the joys of farming. It was very pleasant to be in the pundit's seat for a change. It was also very much a case of the one-eyed man being king.

Next morning I had milked and taken the churns to the top of the lane and, with Shirley and John, had finished most of the early chores before our guests came downstairs. They had slept well. They expressed amazement at how relatively painless it all was. Not even the roar of the tractor under their window had wakened them.

There are, just occasionally, days which stick in the mind. This was one of them. We declared, as far as it was possible, a holiday from work. After lunch we all climbed the mountain, following a winding sheep trail which snaked along the steep flanks, edging steadily upwards, until we struck off through bracken and outcrops of stone and pulled ourselves up to stand breathless, at the peak.

Below us the land stretched away like a multi-coloured, crumpled rug littered with the remnants of children's games. Tiny houses and churches, farms and farm buildings with white corrugated asbestos roofs, toy trees and pencilled roads with silent, miniature cars moving along them.

There was our farm with its house, barn and buildings, and the rocky lane climbing up steeply to the milk stand and the road. Through field glasses we could pick out dairy cows, calves and sheep grazing in the fields. On the same level, to the east, Willem's place and Old Jonathon's red brick farmhouse and buildings could be identified. Above them, right on the crest of the ridge, was the black and white Forge, and away to the west, also higher than Egerton, Howard's farm could be seen. He was burning something in the field by the house, probably old hedge trimmings judging by the way the flames were leaping.

The same binoculars swept round the mountain's wrinkled sides and our guests picked out, excitedly, ponies which they were certain must be wild. So they were for most of the year, until their owners came along to round up the mares and rob them of their foals.

Beyond it all, blue with distance, the hills marched away into Wales. Pot-bellied, sun-darkened clouds hung over them and rainstorms stilt-walked the land, coming threateningly towards us but always sweeping away round our mountain to leave us in sunlight.

This area had been favoured by early peoples. Stone and

bronze tools had been set to work on this land. Nearby were the remains of prehistoric dwellings. It was easy this day to imagine how they too must have stood at this spot and looked out over this same landscape only then, no doubt, it had been wilder, perhaps marsh and forest.

We scrambled down again and trudged home, tired but pleased with ourselves, ravenous from fresh air and exercise. There was the usual work to be done but the extra helpers speeded things and in the evening we drove into town.

Our intention was simply to stroll around, show our visitors some of the sights, and enjoy a quiet drink in one of the old pubs. Not to be! The place was jammed with Morris dancers. There had been a meeting of the Morris Ring, and teams from all over the country had taken part.

The pub we had in mind was an old, half-timbered place. Pilgrims had stopped there for refreshment en route to holy shrines. It was tarted-up now for the tourists but something of the past still clung.

One of the Morris teams was in possession, drinking pints of bitter beer. Their leader had a great repertoire of rather risqué folk songs, and he sang them in a clear, pleasant voice while his fellows stamped in time with the clogs they wore.

Later another team arrived and after some banter, challenged the first to step outside and dance. The whole pub debouched into the street to watch as the teams danced against one another, holding up the traffic and drawing people from neighbouring pubs, until a good-natured policeman arrived and ended it all. Everyone, challengers as well, crowded back into the pub to tackle the bitter again.

Our friends departed the next day for London more than half convinced that this was the normal run of our evenings out. Their whole visit was a tremendous boost for us and for a day or two after they had left the place seemed quite empty.

32 A chauvinistic cockerel and a puppy dog

The acquisition of Chanticleer was just one of those things that happens at sales. There was nothing premeditated about the transaction. Money was too short to allow impulse buys, but he seemed a bargain and my triumphant return home was only slightly marred by Nicholas Paul looking into the car and proclaiming loudly, 'Now he's bought a cockerel!' He made it sound as if they were all continually astounded by the incongruity of my purchases.

The truth is that my original intentions had Chanticleer reclining, nicely browned and stuffed with sage and onion, on a dinner plate and accompanied by appropriate vegetables. At twenty pence he was surely a good buy.

It happened at a dispersal sale for a farmer whose place was situated near the top of the mountain but on the far side from us. He was moving to a nearby Midlands town to take a manual job on the municipal staff. He was a red-faced, cheery individual, born and bred on the place which he had inherited from his father some five years previously.

The farm was about forty acres, but the soil was poor and thin. The grass showed it. Even for a man like him, reared in frugality and hard work, it offered little. But allowing for this, looking at the painstakingly tended garden and the neat, whitewashed cottage which commanded magnificent views, it seemed a daunting leap to move to the anonymity of an industrial town.

The family's finances had depended to a large extent on a flock of hens. About a hundred were being offered in lots of twenty, eagerly snapped up by buyers. But when it came to Chanticleer there was a hang-up. No one wanted him. Farmers' wives bought hens. They saw them in terms of eggs. They already had cockerels running with their flocks. They didn't want another.

He really was a handsome bird with a lot of gamecock in his

breeding. White with streaks of black, bright-eyed, with arching tail-plumes and an aggressive tilt to his head.

'Good stock-bringer,' the auctioneer said. 'What am I offered?'

He was offered nothing. No one spoke.

'Come on,' he said in a slightly exasperated tone, 'a fine bird like this is worth a few shillings.' No one spoke. 'Pence?'

A broad-shouldered man said, 'Five pence.'

This was patently ridiculous. I had this vision of roast chicken. 'Ten pence.'

'Fifteen.'

'Twenty,' I said firmly. My rival knew when he was beaten. He walked away. The auctioneer knocked the bird down to me.

A plump woman said encouragingly, 'He's worth that for the pot.'

'Just what I had in mind,' I said. 'It's not every day you can buy a cockerel for twenty pence.'

Howard came up grinning. 'The last of the big spenders . . .'

I said something rude in reply and we went on to the next sale lot.

Apart from attempting to peck a lump out of his new owner, Chanticleer came quietly. A bit of twine round his legs and he lay on the car floor behind the driver's seat, hardly moving until we arrived home. When I brought him out to show the family, he fluttered his wings and managed to twist his neck in an impossible U-bend to peer at them. 'Quite plump,' I said. 'He'll make a fine meal . . .'

There was an immediate outcry from the kids. 'But it's Chanticleer. You can't eat Chanticleer.'

I could and would. So would John. But we were in the minority.

It appeared that, fortunately for him, the cockerel was the image of the bird hero of a children's story. Even more fortunate for him, they had only just received the book as a present and the pictures were fresh in their minds.

Shirley hovered between playing the Good Fairy who would save Chanticleer from the pot or taking a supporting part to her husband who was now cast in the role of a ghoul. She chose the former. The freezer was full anyhow. 'Eat him?' she demanded indignantly. 'How could you suggest such a thing? He can go in with the hens, at least for the time being.' So, off came the twine and into the henhouse went Chanticleer to introduce himself to fourteen middle-aged spinsters.

'That,' I said, as we locked up, 'is a helluva thing to do to any male.'

'It's better than being eaten; make him earn his keep,' she said righteously.

There was no question about the cockerel's determination to do just that in the only way he knew. Nor was there any doubt about his readiness to accept the challenge presented by his room-mates. There was an absolute bedlam of sound from the henhouse well into the night. Some of the screeches were quite alarming.

When we opened the door the next morning, I expected to find the place filled with feathers. Instead, about half the hens shot out as if the devil were behind them, the rest came out looking ruffled and very thoughtful, and one little brown hen emerged obviously delighted with the whole thing. She followed the newcomer around faithfully, seldom letting him out of sight. Where he went, she went. If they did get separated, even for a minute, she rushed around frantically until she found him again and then stayed as near as possible to her feathered hero.

For his part Chanticleer settled in very well. He took his duties extremely seriously and at length succeeded, at least for long periods of time, in weaning the hens away from their vigil at the kitchen door. He organized their scratchings and led the raids on the vegetable beds. If the cats got too close, he saw them off fearlessly, and when, on still days, the sound of the neighbour's cockerel floated over the ploughed field, he stretched up his neck, filled his lungs and crowed right back

Indeed, he did so well and became so chauvinistic, Shirley began to have second thoughts about saving him. But there it was, the freezer was continually well stocked, you could hardly clobber the cockerel for doing what came naturally to him, and there were always the feelings of the little brown hen to consider.

The day Peter the Puppy arrived was a bright mark on the farm calendar. He was a small boy's dream. A tiny Jack Russell terrier, all busy, bristling energy, ready to tackle anything and everything.

For Nicholas his arrival ended too long days of waiting, during which the sound of a car was sufficient to have him drop whatever he was doing and come running expectantly. Each time it happened and there was still no puppy he grew more dejected.

But, at last, it did happen. A saloon car pulled up and tall, greying Mr Teale, an agricultural adviser, emerged plus one fat, squirming puppy. It was love at first sight. Boy and dog looked at one another and there was an excited yell, 'It is Peter. It really is!'

The explanation, I think, was that 'Peter' was the puppy hero in another children's book which had captivated Nicholas. We had promised him one of his own one day and now here it was, the very image of the one he had drooled about. It was almost too much. He managed, somehow, to gulp some kind of thanks to our visitor and then, plus Vicky who had been reluctantly acknowledged as owner No.2, disappeared to show the puppy round the farm buildings.

I was rather embarrassed about the speed with which they disappeared, but the welcome afforded the puppy was obviously sufficient reward for Mr Teale. He had refused any payment, his own concern being to find the dog a good home.

'They'll look after him,' I promised. It was such a superfluous thing to say, it made him laugh.

His wife had hand-reared the puppy when the bitch went dry. They would have liked to keep him but there were already half-a-dozen other Jack Russells in their cottage home.

To my relief the kids reappeared long enough to wave when our benefactor left. But before his car had disappeared up the lane, kids and puppy were back to their explorations. There is a limit to a puppy's stamina, however, and soon afterwards he seemed to us almost asleep on his feet as he stubbornly followed the pair of them.

So we fed him, settled him in a corner of a well-strawed calf pen and left him for the night: a small, curled-up, black-and-white ball, sleeping soundly. It was his first night in a strange place, so a few whimpers at least were to be expected and could have been forgiven. There was nothing. Not even the smallest yelp.

The explanation was soon evident the following morning. Somehow he had wriggled through the dividing partition to reach the milk calves next door. And there he lay, contentedly, between two of them, warmed by the heat they generated and very, very much at home. Not even when Shirley appeared with calf milk did he wake.

This was the shape of things to come. As he grew, his usual home became a corner in the foodstore where he was supposed to be a deterrent to raiding rodents. But if the nights were cold he simply moved in with the calves and they accepted him without question. When they were fed he came along in the hope of getting his nose into their milk and the calves, once fed, would lower their heads and stand patiently as he licked their milky whiskers and tidied them.

However, there was a flaw. Perfect though Peter might be to his owner, he had a long tail. In the struggle to save him when the bitch's milk had failed, the tail had been forgotten. When it was remembered again Teale's wife could not stand the thought of having it touched because, by then, he had become very special. So she had shelved the decision, but now off it had to come. The mere thought of his puppy being docked was enough to send Nicholas weeping to bed. His big brother explained patiently that terriers were hunters, they chased rabbits and rats into bushes and brush. Long tails could be caught and damaged. Besides, they looked wrong. No, it had to go.

There was a further complication before the dastardly deed was done. Nicholas Paul had a most unexpected ally. It proved when we contacted him, that the vet, of all people, did not relish docking little dogs' tails. 'You might grow to like it,' he said hopefully. 'It's not very long. Well, not what you'd call really long.'

It curled over Peter's back like a broom handle. It had to come off.

'All right, if you are determined,' he said, with the air of Pontius Pilate washing his hands. 'If you really do think it ought to be done, bring him to the surgery next Wednesday.'

None of us fancied breaking the news to Nicholas Paul. It was a tightly kept secret, about the only one I can remember being kept in the family. By the time the boy came home from school it was all over. There was his precious pup with only the merest stump of a tail and even that was stitched and bandaged.

The puppy was perky enough but the boy burst into tears. 'I loved that dog's tail,' he sobbed to his mother.

John and I quickly remembered jobs that had to be done elsewhere and left her to comfort him.

The tears were short-lived. Boy and puppy, bandaged tail and all, were soon back at their games. Just over a week later the stitches came out and the vet and owner both had to agree that it was a big improvement. It was just as well it was done because, as he grew, Peter soon proved an enthusiastic, untiring hunter, always happy to plunge into bush or bracken after some plump rabbit.

33 Running out the rams

Now that the hay had been safely harvested, our friends became sheep-conscious. In a few weeks the rams would be run out to join the ewes which were fat and healthy from the summer feed; turning the rams out about the end of October meant lambs arriving about the end of March, when we might expect some improvement in the weather and some sign of the new grass. Lambs which come in the earlier, colder months need a lot of attention if they are to survive and fatten; in the really cold spells they often freeze to death before they are dry from their birth.

The fat lambs were all gone, even Charles the orphan. Freda had gone into the flock to grow into a breeding ewe. Now, we were told, was the time to weed out the older ewes likely to bring trouble at lambing and replace them with younger stock. The men began to talk about 'going into Wales' to buy the trim breeding ewes the Welsh farmers would be bringing down to the markets from the high farms.

'These days it's all lorries,' Aaron 'from up the mountain' said one evening in the Forge. 'When we was kids it was all to be done on foot. They'd start walking the sheep down from the tall hills before it was light. Sometimes they was days on the road. They'd rest the sheep at farms along the way and move on the next morning. The bloody lanes would be choked but there wasn't all these cars about then. You come on them and you'd to wait and let them go and no good shouting the odds. Them old Welsh shepherds would sooner fight than talk.'

'Some of them could hardly talk a word anyway,' Phil Morgan, a lean, dark-haired farmer with a slight stoop, said. 'Many of them had only the old language. Just a bit of English but always enough to understand if you offered them a beer or tried to do them out of a shilling. Nowadays they all speaks English and there's not much of the old tongue heard about.'

'Television,' someone said.

Aaron pulled at his beer and sucked his yellow teeth. 'The men was hardier and the sheep was hardier. There wasn't all these cure-alls to keep the weak uns going. The poor uns died and the strong came through.'

Old Jonathon would have nothing of it. 'The sheep's all right now, Aaron. I don't know so much about the men. Sometimes I think we'm the last of our kind.'

'Thank the Lord for that, then,' Phil Morgan said and went to get another round.

Somehow, as usual, although nothing seemed to have happened, it was agreed. About half-a-dozen of them, headed by

Phil Morgan and Aaron, would go into mid-Wales to buy for themselves and for friends. I was invited to join them but there was neither the time, nor the money, to spare.

A few days later Howard rang to ask if I was interested in buying six ewes he had got in a trade.

'Good strong Cluns,' he said 'from the top of the hill. Come and look at them. If they don't suit, you don't have to buy. '

We went in the van and found him working round the buildings grinding some of his own barley for cattle feed.

'They'm in here,' he said, leading us to a small, wooden-railed pen. 'Nothing wrong with them. Some's two years old, others more and there's one that's only a yearling. If I hadn't mentioned there was six, I might have kept the little one for myself. But six I said, and six you can take if they suit.'

They were strong, lean ewes with big frames. On the better grass of our farm compared with the hill pastures they had just left, they would soon put on flesh.

'I'll take them. How much?'

'No, no,' he protested. I was not following protocol. 'Look at them first. Look at their mouths.' He caught one and then another and pulled their mouths open to show their teeth. 'Their udders is good, nothing wrong there.'

I was unrepentant. 'Your word's good enough for me. How much?'

'Well, say £47 the lot? That's giving me a couple of pounds on the deal. No more. They're worth it. You'd not get this lot in the auction for that.'

'I'll take them.'

He pulled a knife out of his trouser pocket and opened it. 'You ought to have looked at their feet too. It happens there's nothing wrong with these but always look at the feet: Footrot.' He caught a ewe and sat it on its stern and proceeded to trim its feet. It took him only a few minutes to tidy the feet of all of them and he shut the knife and dropped it back into his jacket pocket.

'I'll write a cheque,' I said and we trooped into the kitchen to scrounge tea from his wife.

'You ought never to take the first figure asked,' he cautioned.

'Always ask more than you expect to get, then you can come down a bit.'

The tea was hot and sweet.

'Mind, if you hadn't said outright that you'd take them, I might have put a bit more on the asking price. They're worth more than you've given and that's straight. There might have been a bit more in it for me.' A dreadful thought struck him: 'You didn't think like that, did you?'

I wrote the cheque.

'Damn,' he said. 'I think you might have. You're getting too damn fly. Did you really think like that?'

I knew better than to confirm his suspicions. 'You wouldn't do a friend, Howard. I know that. You're too generous. Everyone knows you wouldn't do a friend.'

He took the cheque and folded it. 'Now don't you go telling people what you got them for. It's stealing. There's men in gaol for less. Make it a bit secret, like.'

'Not a word,' I said and turned the talk to rams.

'Edart's got some ready to go,' he said. 'You can't do better than go to him.'

So we returned home with six ewes in the van and an arrangment to look at Edart's rams.

Edart was the bailiff on a big farm which stretched away up to the common land on the mountain. Locally it was conceded that he 'knows a bit about sheep'. Praise indeed. We often saw him riding a roan shepherding pony with a couple of collies trotting along behind, on his way to check on the welfare of the hundreds of sheep which were his responsibility.

When we arrived there were a half-dozen stocky Suffolk rams on show in a covered pen and about as many farmers looking them over. The man himself was there, chatting quietly as he always did, explaining that they were all guaranteed sound and 'ready for work'.

'There's not much to choose between them,' he said, indicating the animals. 'They'm all the same price to you lot: £20. I can't let them go for less. There's nothing between them.' He ran them round for our benefit.

The men already there were discussing the good and bad

points of the six – all yearlings – and occasionally probing into the even-woolled fleeces. For their part the rams kept together and looked suspiciously at the humans leaning on the rails.

The others were handicapped by knowledge. I went by appearance. It seemed to me that in some subtle way one of the rams was more compact and even than its companions. A light mark on its forehead distinguished it. I needed only seconds to decide. 'That'll do for me,' I announced. 'I'll take that one.'

Apart from acting in unseemly haste, since such matters should be deliberated at length, it appeared that I had also snatched the cherry from the top of the trifle. The other would-be buyers turned shocked, accusing faces towards me.

'I had about decided on that one,' Frank Johns, a local farmer, said.

Edart grinned. 'Gone. He's bought it.'

'We were still talking, I didn't know you were selling,' Frank persisted.

A gesture seemed called for from me. 'Let's toss for it.'

Edart spun a coin. Frank called 'Heads'. It came down tails. The ram was ours.

'I don't know,' the loser grumbled good-naturedly. 'You'm supposed to know nothing about sheep and here you come and get the best of the bunch.'

Edart was enjoying it. 'Books. He reads it all in books.'

The difference, of course, was marginal. All the rams were sold and everyone was satisfied – particularly us.

Settling up, Edart said, 'I can't knock anything off him, Jacky. But if you like, I'll keep him here until it's time to turn him loose.'

It was a big help. The ram needed to be kept separate from the flock until the appointed time. We had nowhere, no small paddock, that would serve, so it would mean setting aside a whole field. Besides, there was always trouble with the ewes trying to break in or the ram trying to break out. It was really a very useful offer and we accepted gratefully. Altogether a good day. We went home very content.

We collected him four weeks later on the last Friday in October. This was a generally accepted date locally to loose the rams and it was advisable to tie in with neighbours because then you could swap after three weeks. As ewes 'come on' at three-week intervals if they are not in lamb, changing rams meant that any ewe not with lamb would be served by a fresh male. I did hear of a case where the ram had been infertile (it could have been because of an unfortunate knock) and the ewes it served did not take; the farmer did not change rams and there were no lambs in the spring.

Our ram was penned and waiting when we arrived at Edart's place. 'There he is,' the man said with the air of an impresario. 'The pick of the bunch, fit and pawing at the ground, ready to go. He hasn't been used.'

The animal was in prime condition and easy to load because he had always been quietly handled. 'Sixty-odd ewes?' Edart said as we slammed the van doors shut. 'That's nothing to a young ram like this. About a fortnight's work, I'd say. Bring him back if there's anything wrong. When you need a swap, let me know.'

Introducing rams calls for caution. Where there is already a ram with the flock he will be jealous of his ewes and will fight intruders, sometimes to the death, for them. A common local practice was to put the two rams together in a small pen which did not give enough room for a full-blooded charge. It forced them to accept one another, and once they had settled down they could be turned out together. In another method, farmers simply yoked rams side by side until there was no conflict.

There were no such complications for us. We had brought the ewes up to the Calf Field the previous night and we just drove the van into the field with them. The ewes were rutting. They scented the ram and closed round the van excitedly.

Before being released the ram had to be raddled. This meant smearing his chest with a blue paint-powder mixture which would rub off on the ewes he served. After one week we would use a different-coloured raddle and this would indicate which ewes were likely to lamb first.

When we did open the doors the ram stood for a long moment looking down at the milling ewes. He was the missing link in the breeding chain. He jumped down, the flock closed round him, and the whole lot went off, back to the lower fields. It was as natural as the seasons.

Next day a quick check showed that nine ewes had been covered.

 34 The changing seasons 

This early morning the red fox was walking along the top lane carrying a fat rabbit, so complacent and pleased with himself, he never heard the tractor until I was almost on him. Then he burst through the hedge and ran away for dear life down the dingles.

It seemed to me that he was carrying summer in his mouth. The season ran after him. The sun was draining the goodness from the grass, and we were now having to supplement the cows' grazing with increasing quantities of concentrates to maintain the milk yield.

The same sun ripened the blackberries and the kids went round the hedgerows picking the fruit into baskets and jars for Shirley to turn into jam or portion out into packets for the deep freeze. They had competition because the two sows insisted on accompanying them and delicately picked and ate the berries from low-growing brambles.

There was a harvest festival atmosphere about the area. Price came down the lane and combined our heavy-headed barley for us. We got 100 bags of grain in four hectic, back-breaking hours and lugged it on our little sack truck into the bottom end of the foodstore to be used for winter feed.

It was a scramble to get everything done, but we were growing used to that and, somehow, we coped. Howard loaned us a tractor and his long trailer; with this and our own we lugged nearly one thousand bales of beautiful, clean, golden barley straw from a neighbouring farm in three days. The cost was one shilling a bale and we built it into a square rick along with the straw from our own barley crop and covered it with black polythene sheeting to protect it from the rain and snows that winter must bring.

Crab apples ripened but got no sweeter. The kids willingly climbed the wild trees and picked basketfuls for their mother to transform, by some kitchen alchemy, into luscious jelly. The bitterness was no deterrent for the sows. They rooted about under the trees for windfalls and, as the kids clambered about, they waited hopefully for any crabs that might be shaken down.

One Monday Shirley and I travelled to market and bought plums – Yellow Eggs, Czars and big, juicy but expensive Victorias – to be bottled and put away in the cupboard I had built in the utility room. We got apples too, some of which she pulped and froze, others which we wrapped in newspaper and stored away in shallow trays for the winter.

A small van arrived one morning and delivered twenty-six cheeping chicks. They were Cobs, intended to reach edible size in time for the Christmas trade. At this stage they were fragile balls of fluff needing to be coddled under the infra-red lamps and protected from just about everything if they were to survive. The kids drooled over them and I had to warn, 'Don't get too fond of them because they are going into the deep freeze to be eaten.'

This time there was an unexpected response. Nicholas Paul smacked his lips in anticipation and declared, 'Lovely. With roast potatoes and brussels.'

Slowly, almost imperceptibly, but remorselessly, the countryside was changing. Summer had merged into autumn. On the mountain the bracken was dying, silver-leaved foxgloves stood eyeless against the hedgerows, in the lane the willow herb was a tangled ruin.

The kids kept bringing in offerings of coloured leaves, twigs, rushes and grasses for Shirley to put into vases. The house began to look like a jungle clearing and in the end she had to say, 'No more or we shall get lost in the undergrowth.'

There was a slow-down in work. For almost the first time since we arrived, we could take a long look at the farm and at ourselves. The need now, our friends took pains to remind us, was to prepare for the oncoming winter. It was like getting ready for a seige. 'Any little jobs needed round the place, you'd better get at them now afore the weather comes down on us,' Griff advised. 'You walk round and see what needs to be done.'

There was no need to walk far. There were so many jobs to be tackled, I typed out a list and tagged it on the inside of the foodstore door. Pens needed to be mucked out, cleaned and disinfected, ready for stock which would have to be housed during the hardest months. There was woodwork to be repaired and I lost a fingernail making the first of two heavy pen doors.

Matthew, Old Jonathon's back-aching brother, looked at my bandaged finger and said, 'You keep on chipping bits off yourself, Jacky, there'll come a time when there's nothing left.'

Above all, though, there were roofs to be made good against the weather. They had been much neglected, particularly on the older, traditional buildings. There were some very bad patches on the foodstore roof: in one corner rain poured in but, fortunately, did not fall on any of the stores.

It was some help that the foodstore roof could be reached from the flat top of the collecting yard. I used our London house ladder to climb up there and then dragged it up behind me to work on the roof. Just why repairs had been neglected was easy to see. Many of the joists were broken or bellied down with the weight of the tiles. Battens were broken, many had slipped and a lot were missing. It really needed a fullscale reroofing job but cost made that out of the question: ours had to be a patch and improvise exercise.

Even this was far from simple and it was certainly not very pleasant. It necessitated placing the ladder across the steeply sloping roof to spread my weight. I crawled along it very gingerly, trying to keep movement to an absolute minimum. If I shifted my position too quickly, there was a danger of the ladder sliding. It did, several times.

Lying face down over rotten timbers, juggling with tiles and tools two storeys up was not my idea of fun. To make it worse, the first storey floor had been taken out and there was nothing between me and the concrete at ground level. If a tile slid through a gap, and a number did, it seemed to go on falling for minutes before shattering below. Every time I tried to hammer in a nail, all the tiles in the immediate area bounced and threatened to be shaken loose. There were several minor tile-slides but, gradually, the biggest gaps were closed and in the end, if I did not stop all the rain coming in, at least it was reduced to drips.

One moment of light relief came when Dorfie Pig found that the foodstore door was unlatched and nosed it open. She was intent on a spot of pillage but mistrusted her luck. It was a little too good to be true: she smelled a trap and hesitated before stepping inside.

A look of shocked amazement spread over her face when I bellowed, 'Get out of there.' The sound reverberated from the thick stone walls, making it difficult for her to determine where it had come from. She snorted and spun right round like a top to see where I was, but, for some reason, never thought to look up. I shouted again, louder; the noise was impressive, magnified in the confined space. It bombarded her, and it was too much. She turned and bolted out of the store and way up the stockyard as if expecting at any moment to be set upon by butchers.

My yells disturbed the sparrows which had made the foodstore their home. They quit hurriedly via the paneless windows, and it suddenly occurred to me that the swallows, swifts and the little house martins had all left us. I had seen them congregating on the telephone wires a week or so before but had not consciously missed them until now.

'They've gone back to Africa,' Nicholas hastened to inform me. 'They'll be flying over France or Spain now.' It appeared that they had been 'doing' the subject of bird migration at school and he was never reluctant to help with my education.

The departure of the swallows made the end of summer official. It was a hard blow. We are a warm-weather family. Shirley's idea of Hades is a hard winter. She was already busy knitting with wool nearly as thick as her fingers, taking everyone's sizes and tucking the end products away in various cupboards, ready for the onslaught.

Another step in our preparations for the winter siege was taken on the first Thursday in November when the mobile grinder arrived to convert our home-grown barley into animal foodstuffs. These enormous, self-contained vehicles were relatively new in that part of the world. Previously farmers had either ground their own barley or taken it to the local mills. Many continued to do this, but for farmers like us without the necessary equipment, the service was a great time-and labour-saver.

For this first visit we tipped the bags of barley grain into a large galvanized bath. From here it was sucked through an extendable vacuum pipe into the grinding and mixing mill. (The merchant provided various additions for the mix including wheat and molasses.)

The finished product emerged from the machine slightly moist and sweet-smelling. We bagged it as it came out and trolleyed it into the store for keeping. The stock loved it and we ended with three tons produced at less than half the cost of commercially marketed feeds.

The operator was a tall, tow-haired young man from East Anglia who had been brought in to start the service and train local people. He was homesick for his family and the flat fenlands. 'There's too many hills round here to suit me,' he told us. 'I feel closed in, choked-like. It's all right for folks that's always lived here, I suppose, but not for me. It's too far from the sea. Where I come from you climb a ladder and see for miles.'

He opened the machine and borrowed a broom and shovel to clear out the milling compartment. It produced nearly two sackfuls of finely ground meal. 'You take it,' he said. 'It's marvellous for geese. They rear a lot of geese where I come from.'

We accepted the gift but saw no reason to waste it on bellicose old Moses or his waddling spouse: they were fat enough. We preferred to mix it with any waste milk that might become available and feed it to the twenty-four piglets which were weaned now and doing very well. The hog pigs in both litters had been castrated and the aim was to take them all through to porker weight, between six and seven score, before selling.

(The two sows had gone back to the boar within a few days of their litters being weaned.)

Welcome though the home-ground meal was, the mobile mill's visit had made some hard work. The bags of grain and the finished meal were heavy and difficult to handle. So when we heard of some old houses nearby being demolished, we went along and bought the floorboards that were being ripped up. With these we refloored the granary which was over one end of the foodstore and broke a small window through the wall overlooking the stockyard. It took the backache out of the work. The next time the grinder came, the grain could simply be sucked into the mill through one tube and the finished meal blown via a second tube through the window into the granary and stored there. We then brought it down in buckets as needed.

Life other than ours went on round the farm. Foxes were mating. We heard the vixens crying in the long gulley and, in her agitation, one of them came right into the farmstead. She went off again very hurriedly with the collie and an ambitious terrier puppy hard after her but soon left behind.

On another occasion, walking with the collie, I came on two foxes engaged in their courtship ritual. They were so intent on mating that we came to within twenty yards before being noticed.

35 The blind colt at the pony sale

Everyone appeared to turn out for the pony sale at the Forge market. There were two a year, and people insisted on calling them 'pony sales' although the catalogue preferred to describe them as 'A Sale of Riding Horses, Ponies, Young Stock, etc.' Perhaps the popular name was recognition of the fact that every kid for miles around seemed to have met there hours before the start. The catalogue might well refer to entries as being 'thoroughbred hunter type' with impressive and impeccable breeding, but the kids were there for the 'ponies' and could talk about nothing else.

Gangs of little boys in patched trousers racketed round the place with mongrel terriers racing alongside. Little girls in jeans and handed-down, too-big Wellingtons dragged smaller children by the hand and threatened to 'tell Mum all about it'. Bigger children walked more sedately and affected a casualness belied by the brightness of their eyes.

The atmosphere was completely different from the cattle and sheep sales which took place regularly in the same ring. There were 130 horses and ponies on offer singly or in lots. Most of the familiar local faces were present but it was 'outsiders' who did most of the buying.

There was something of the carnival that attended the old horse fairs. In a field adjoining the auction, stalls had been set up to sell old and new horse harness: bridles and saddles, rope halters and a variety of leatherware. Jodhpurs of all sizes, riding boots and whips were on offer. You could buy your daughter a hard hat to protect her head from breaking should she fall from the family pony. The briskest business was at a stall offering coloured rosettes to decorate your pony or, perhaps, to hang round the loosebox to impress your friends.

Most farms, if there be the mind, can find a corner to tuck away a pony. On many around us there were ponies as precious as family heirlooms, handed down from child to child, domesticated, loved, trusted and so experienced that they were virtually four-legged nursemaids. In one perfect example on a neighbouring farm a fat mattress of a pony was shared by five children. The youngest, not more than three years old, spent hours sitting astride the pony's broad back doing nothing, simply sitting with a happy smile on her dumpling face. One day I passed them on my way to town and they were still there when I returned over two hours later: in that time they had moved about twenty yards. The pony had taken the initiative to reach some fresh grass. The child had the same rapt expression of dreamlike content.

Our kids had been at the sale an hour before we arrived. In one corner pen two little grey fillies were being petted by a cluster of kids. Vicky was one of them and she had already lost her heart. No farm, she explained, could ever be complete without a grey filly pony. Escape came in the form of Nicholas Paul who dragged us along to see a grey muzzled, 'outgrown' pony in another pen. Such an animal, he said, was far superior to any grey filly.

Details of the sale entries were given in the catalogue when there was anything to give. Some were just described as 'Welsh Mountain colt foal' and there were some thirty lots entered by hill farmers as simply 'Foal – details at time of sale.'

There were big, muscular roans and blacks and chestnuts and bays said to be experienced in the hunting field. A palomino was one of a score described as 'a good children's pony'. There were animals guaranteed 'quiet in traffic' (a useful quality anywhere), ponies described as 'brood mares' and others in foal to Arab stallions. There were animals of thoroughbred lineage with successes achieved in gymkhanas, show rings or over tough, cross-country courses.

Some of the foals were sold as 'suckers' which meant they were still suckling the mare. Local farmers would sometimes buy such a one and rear it with the milk calves. They were fed, housed and fussed with the calves and the end result was an animal more than half-broken before it had even tasted a bit or had a saddle thrown across its back.

Many ponies came on their own four feet to the sale, puffing along under owners who had outgrown them and needed something bigger and stronger to carry them along the local paths. The sellers were often plump-thighed country lasses who might at some time have aspired to great deeds in the show jumping rings. Now all they wanted was something they could ride knee-to-knee with the long-legged, lean young men who strode about the place dispensing instant judgements.

When the cattle trucks arrived with stock from the outlying farms there was a rush of children to watch the wild-eyed ponies unloaded. These were no suburban riding-school pets, not yet at any rate. They had been born on brackened commons and known little restraint in their roamings. Now they were confronted with the sights and sounds of the sale and they snorted and tossed their heads as they came down the sloping tailboards. In the pens they crowded into the corners furthest from the gesticulating bipeds who perched on the top rails or leaned against the gates and assessed their merits and demerits. These were animals brought for the attention of the professional horse-breakers who came to such sales looking to buy for fifty pounds an animal which could be transformed into a mount worth six times that much in the wealthy South-East.

There were three of these professionals at the sale this time, and between them they took all the wild ponies. One of them, a bowler-hatted gent, seemed to bid on everything. He had three hard-looking acolytes, probably sons, in support and they organized the handling and transport of his purchases. They gave loud appraisals of each buy and let everyone around know that their trade was horse-breaking.

They showed their skill when eight hill ponies panicked in the ring and crashed against the barrier. The top rail snapped and the ponies stampeded through the gap and scattered the crowd. The jagged wood struck one bystander – no one we knew – giving him a nasty gash on the forehead. Someone produced a wadded handkerchief and he stood cursing the animals and holding the cloth against the wound to staunch the bleeding.

The acolytes soon brought the ponies under control and penned them safely. One of the runaways – a rough-coated piebald – half-heartedly reared at one of the young men. It was scared rather than aggressive and the acolyte was out of distance and in no danger, but the action angered him. 'I'll break that bastard,' he told his companions. 'He's to be mine.'

For everyone except the injured man, the incident had been an exciting interlude. It took people a few minutes to settle down. A length of wagon rope was brought and laced across the gap and the sale continued.

The conversation piece of the sale was a well-formed piebald colt foal which was penned with its dam and three other ponies.

Most of the time its face brushed the mare's coat. When she moved the colt went too. If the foal strayed a yard or two away the mare looked for it and re-established contact. They were so practised and natural it was hard to believe that the colt had been born blind. Now it was, perhaps, nine months old and a fair size.

It was listed separately from the dam and sold on its own. Whereas other ponies trotted round the ring or milled together in the centre, the blind colt stood quietly alert, perhaps listening for the mare to call.

The auctioneer was a horse lover and distressed at the sight. The colt should have been 'put down' immediately it was seen to be blind, he said. It should not have been reared. It would have to go straight to the knacker's yard. The owner was embarrassed and muttered that the pony had been reared on the common, running semi-wild and the blindness had not been noticed.

Surprisingly, when bids were invited there was brisk competition between two of the horse-traders. It was knocked down for £30 to the bowler-hatted one with the final warning, 'Remember it must be destroyed.'

'Oh, it'll be knocked down all right,' the farmers round us said grinning. 'It'll be over the Channel to feed them Frenchmen. There's good money in horsemeat and a young 'un like that'll be as tender as any calf.'

When it was taken from the ring the colt walked easily through the gate, but in the aisle leading to the pens it stopped and called anxiously for its dam. The mare whinnied in reply and the colt would have turned towards her had there not been a fence in the way and men's hands to guide it to the pen where it must wait to be collected. The separation was final. It called once more but the mare herself was in the sale ring and there was no response. After this the colt accepted its fate and stood mute with its shoulder touching a heavy corner post as if it drew comfort from the solidity.

Before the sale we had no intention of buying, but, impulsively, we decided to bid on a dainty palomino sucker. Fortunately something distracted my attention at a crucial point and the foal was knocked down to someone else for sixteen guineas. It was a bitter moment for Vicky but something of a relief for Shirley and me.

When the sale was over we estimated that the 'wild' ponies had averaged between £50 and £60 each. Not all the stock went as cheaply. There were horses which fetched upwards of £200 and one beautifully groomed dappled grey whose equally well-groomed owner disdained a bid of £300 with a perceptible twitch of her well-shaped nostril.

We stayed talking with our friends, Howard and Dilys, and went with them to sit outside the pub with shandies before returning home. After they left I walked across the road for a final look round. Most of the stock had been taken. Two cattle lorries loaded with ponies were parked just off the road and the drivers stood by them chatting. In one pen three ponies stood together and four more waited alone. The pen which had held the blind pony was empty and the gate was swinging loose. I walked down and closed it and then went back to Shirley and the kids. The sale was really the end of autumn.

36 Egerton's skating ducks and drain problems

The talk now was how long the land would stay open. How long it would be possible to keep stock outside as winter took over the scene. The sooner they had to be brought in, the sooner they would eat into the stores of hay and barley straw, the sooner the hard slog of the winter routine would begin.

All around there were signs of change. Nights were colder. There was a bite in the early morning air. Trees were laid bare as leaves blew away in winds which came increasingly from the east or north. In the market cafeteria on Mondays the rustic Cassandras supped tea and proclaimed, 'It'll be early and hard this year. Look at all the berries on the hedges. What does it mean? It's a sign of what's to come . . .'

One morning towards the end of November, we woke to find the farm frozen over. The pond looked like a skating rink. Trees were draped with icicles and the grass was white with rime. Puddles and mud wallows were solid.

A month previously Shirley had bought a dozen ducks at seven shillings each from a friend. Now they were more bemused than us by the frost. When they stepped on the pond their feet skidded from under them and, much to the kids' delight, they slid about on their bottoms. At the edge, where the ice was thin and brittle, they broke through and plunged about frantically, feeling for firmer footing. In the end they gave up and pottered about unhappily among the rushes until the kids took pity on them and smashed the ice with some long poles.

The cows were very frisky. The cold snap stimulated them. They had slept in a hollow in the field and been protected from the wind by a line of trees. When they stood up they left an oval of frost-free grass. Their breath was white against the chilled air and steam rose from their warm undersides. It was all very picturesque but sadly, the true effect of the cold weather was to be seen in the milk yield, which was markedly down.

They went out again after being milked and we dropped hay to them in the field but brought them into the covered yard that night. A week later it snowed and they had to be kept in full time.

There were still a lot of jobs unticked on the foodstore list, but one we had managed to complete was to extend the hay racks in the big yard. We bought lengths of heavy wire mesh and built timber frames. There were sixteen dairy cows now and it was essential to give them all easy access to the feed or the more dominant ones would gorge themselves while keeping the others away. Now, if a cow was bullied off one spot, it could simply move to another. As Ellis, my instructor in the art of dairying, put it, 'Fair shares for all.'

Really, we had not done too badly. Most of the important jobs had been tackled, if not to complete satisfaction. The pigs were snug and warm. We had insulated the roofs of their sleeping areas with straw bales, and inside the heat from their bodies worked up a high temperature and a fair old fug.

Alice Capone and her mob, the biggest calves, were still outside and had to be supplied daily with hay. John and I patched two old, free-standing hay feeders that some previous owner of Egerton had left behind, and put them out so that the fodder could be kept off the ground. The calves also shared a bucket of concentrates daily and looked very strong and healthy.

Snow – providing it was not too deep – and hard weather did not worry the sheep. They simply scraped the grass clear and carried on grazing. We put out hay in a long, mobile rack I had bought at a sale, and later in the winter, when things got really hard, they also got a small ration of concentrates.

As winter settled in we found that snow and frost were not our worst problems. If the ground was frozen and not likely to cut up, the cows could go out for a few hours. After the confinement of the yard, they relished the change. There might not be much food value in the grass but they pulled happily at it and benefited from the exercise and freedom. No, our worries really began when it thawed. Mud everywhere! It squelched underfoot and oozed over the top of shoes, boots and Wellingtons. Any movement was hard work. Pushing the wheelbarrow loaded with sloppy muck scraped from the collecting yard and needing to be dumped on the dung heap was particularly hazardous. The routine chore became an obstacle course with aromatic penalties.

The worst never happened but there were several near-misses. On one occasion, struggling to prevent the wheelbarrow from overturning, I could do nothing as it slopped over at one corner into my Wellington boot. Had I moved my foot, the whole load would have gone. It was an explanation which Shirley, who had not accepted my philosophy towards farmyard pongs, found hard to accept.

On a second occasion I avoided the ultimate disaster only at the expense of getting my hand, arm, sweater sleeve and all, as far as the elbow, immersed in the mess.

A thaw also meant that the drains feeding our pond bubbled and burped with water coming down from land higher than ours. As a result the pond overflowed and there was a sizeable stream running down the stockyard. When it froze at night, and it always seemed to, we found ourselves doing a duck act while carrying buckets of calf milk or struggling with 70 lb bags of foodstuffs.

Something would have to be done before a leg got broken. When the weather brightened a little, Thomas, Ellis the cowman's son-in-law, and I – John was back at school – set about checking the six-inch field drain serving the pond. Something was reducing the flow of water along it; the pipe had to be traced back from where it emptied in the bottom field pools to where it began in the pond. This really was back-breaking work. In three places wet patches of ground indicated that tiles were broken or out of alignment. Each time it entailed digging down nearly three feet and working in icy sludge.

'I'll bet you wishes you was back in that nice warm office now,' Thomas said, grunting with effort as he took his turn with the spade.

'Not me, I love it here,' I lied. My trousers were muddied right up to the knees.

He climbed out and handed me the spade. 'You must be bloody mad then. I'd close our farm down over winter if it was possible.'

Each time we repaired the pipe there was an improvement, but the main blockage remained. It had to be somewhere near the pond. We made a calculated guess and picked a spot a few feet from the intake. Luck was with us and we hit the very tile that was giving the trouble.

Over the years the roots of a big elder bush had infiltrated the tile joints and grown inside the pipe until they formed a dense beard. Muck carried into the drain accumulated on it and further reduced the flow. A rod pushed in from the pond end simply passed through the 'beard', which closed again behind as the rod was withdrawn. The bush had to be removed. We looped a tow chain round it and yanked it out with the tractor. It resisted and creaked and brought a fair portion of the bank with it, but it came.

My neighbour Willem did not approve. He had a bit of a

thing about elder. 'Listen to them groan if you hurts them,' he said. 'They'm graveyard trees. You upset them, you'll never lie easy in your coffin.'

I was prepared to take the risk.

When we had replaced the tile, Thomas pulled away the boards we had used to block the drain intake. The result was satisfyingly dramatic. A good head of water had built up. It raced through the pipe at great speed, carrying everything with it, and shot out into the ponds three fields below with a tremendous rush and a great mushrooming cloud of mud. Our pond threatened to disappear down the pipe. So, to control the water level, we had to build a little horseshoe dam a few bricks high around the intake. By outside standards the whole job was paltry but probably the only creatures who might have got as much satisfaction from it as I did were the frogs who took up residence inside the dam when spring came round again.

December proved a pig of a month. It seemed to be all ice, snow and rain. It was bitterly cold and I was continually getting wet. My back played up and Shirley got stroppy and demanded that I see a doctor. The mere thought of facing that female dragon made me worse. There had to be an alternative.

We compromised. I brought in an infra-red lamp and rigged it up in the bedroom. Lying under it with the heat on the seized-up portion of spine helped a lot.

Too many of the locals suffered from this sort of complaint for me to expect much sympathy. 'Farmer's back' was the name they gave it. 'You and me shall have to get corsets,' Matthew, Old Jonathon's arthritic brother, laughed when he saw me come into the Forge.

The effect of the cold weather on diesel engines added to our woes. Some mornings the tractor resolutely refused to start. Old Lil was also unhappy but could usually be bribed with a whiff of Quick Start. I bought a battery charger for £10 and every night plugged it into one or other of the vehicles' batteries, but even this was not always sufficient.

The tractor chose a particularly unpleasant morning to conk out going up the lane with the milk. Fortunately Shirley was taking the kids to the school bus in the 1800; she dropped them off, raced back and we managed, somehow, to transfer the churns and reach the stand before the milk lorry. 'There's trouble right along the route,' Jock the driver said. 'There's some that haven't managed to get to the road. I've left them fresh churns and we'll pick up the milk tomorrow.'

It was something to know there were others worse off than us.

Price, from the top of the lane, helped me to strip and clean the tractor feed system of its rust particles on that occasion, but the following day the same thing happened; this time we towed the tractor into Griff's garage and used his tools to effect a permanent cure.

It was as well we did. A few days later the car's clutch gave out: probably the effect of coping with the ice-bound lane. I had to tow it behind the tractor for eight miles to the village garage and by the time we arrived there was no feeling in my face.

Then, just when we were all feeling very 'down' and sorry for ourselves, Gaffer, the herd leader, produced a healthy, red bull calf. It was a marvellous tonic.

37 Hunting the red fox

The cold weather seemed to stimulate the basic instincts. The Hunt went over our land, lickety split, after the red fox: the hunted first, hounds five minutes later, and ten minutes behind them, the riders, tired and frustrated, out since early morning and nothing to show for it. Nor was there anything to come this time. The red fox was a professional. He had been hunted before. He ran easily across the twelve-acre and paused for a drink at the stream before turning right and running through the water, downstream towards neighbour Willem's territory.

There he had the effrontery to kill one of Willem's fat ducks and lie on the slope of the field under the house enjoying the meal while the hunt rampaged round him. He was in no danger. When they reached the stream the hounds lost the scent. The water had carried it away. They turned left, away from the fox, and hunted upstream towards Old Jonathon's place. A huntsman was posted on the road to intercept them but they had been neatly fooled, and after an hour plunging about the area the whole fandango moved east towards the low hills. As for the fox, he simply finished his meal and trotted home to the dingles. Just as well because Willem found the remains of the duck and began hunting the culprit with a 12-bore tucked under his arm.

Since he knew nothing of this second threat to his well-being, the red fox was unimpressed. He returned the following night and snatched a guinea fowl from the compound close to the farmhouse and carried it away while Willem's dogs barked and rattled their chains helplessly.

It was amusing, in view of the campaign to stop foxhunting, to sit in the Forge and listen to the locals complain about the Hunt's inefficiency. Even Old Jonathon, a staunch supporter for long years, had to agree they did not hunt the area often enough.

The normally placid Willem was incensed with his losses. 'They don't kill foxes, they only plays at it,' he said. 'They upsets the foxes and chases them about and it takes weeks for them to settle down again.'

There was something in this. Several strange foxes were to be seen round Egerton after the Hunt's visit. The day after, I heard someone or something chattering on the other side of a hedge and looked over, expecting to see strangers. Instead there was a fox trotting along and 'talking to itself'. It looked up and saw me but took no notice and simply continued on its way, presumably returning to its own territory.

'It'd pay us to get together one weekend and do the job ourselves,' Willem said. 'Get the terriers on the job and shoot them out. Between us we knows just about every fox earth in the district.'

Old Jonathon would have nothing to do with it. 'I'm sorry about your birds, Willem, but I like to see a fox or two round the place. They belongs here, somehow.'

'Vermin,' Willem said unrepentantly.

'There's a fair few foxes up in the bracken on the hill,' Aaron put in. 'You'd never get them out with any terrier. Not that the Hunt is likely to get them either, they just loses the hounds. You know, I've seen 'em roll in fertilizer to kill their scent and throw off the hounds: it worked too.'

The end was that Old Jonathon undertook to let the Hunt know that some people were not happy with their efforts in our area and thought they ought to do more.

'If it wasn't for the Hunt, we'd soon do away with the foxes,' Willem said gloomily.

He was probably right.

Foxes were not the only quarry in the countryside at that season. Like many of the younger generation, John and his chum, Harvey, were busy rabbiting. Harvey had a couple of ferrets: nippy, brindle hunters. They travelled in a small box strapped to the pillion of his motor-scooter.

He was a brash but likable eighteen-year-old with a thick Birmingham accent, a successful barrow-boy type, long hair and an impressive optimism. Invariably he turned up wearing an 'office suit' and shoes, and shamelessly borrowed my over-trousers and Wellingtons. One day I found an old pair of Wellingtons in a shed. They were size 10, far too big, but that mattered not at all, he slopped round quite happily in them.

The pair of them, aided and abetted by Peter, the terrier, worked zealously round the farm, dropping their nets over the rabbit holes and sending down the ferrets to drive out the occupants.

Their bag seldom justified the long hours, discomfort and effort they put into the chase. Myxomatosis had cut the rabbit population to a fraction of what it had once been. Animals which survived were usually bush dwellers and did not go underground, where fleas carried the disease from the stricken rabbits to fresh victims.

Bush rabbits were more likely to be endangered by the 'rough shoots' which were very much part of the winter farming scene. Round us the smaller farmers tended to pool their land and walk it together. This gave a better chance of making contact with the quarry; often the first bang of a gun would send the game scurrying for quieter neighbouring land where it could not be followed.

Not that the sportsmen worried too much about that. Most of the walks ended up at the Forge for an exchange of hard luck stories. It was noticeable how the pace of the drive quickened when opening time approached. The only trophies I ever saw were wood pigeons. There were three of them tied together at the head, and every one of the party claimed to have brought them down. The argument was how to divide the bag. 'Why not take a feather each?' Griff asked.

In the midst of all this hunting fervour, Vicky and some of her friends returned from roaming round the bottom fields with a handsomely feathered duck and a graphic story of how they had stalked and caught it. There was something suspicious about the contented way it lay in our daughter's arms.

Momentary thoughts of duck-and-orange dinners disappeared when I found its wings had been clipped. It was a decoy duck placed on the bottom pools by our neighbours to lure its wild fellows within range of the guns. It appeared to be a very lonely little duck which, perhaps, did not say much for its efficiency in attracting company. When, under some protest, the girls returned it to the water the bird tried so hard to stay with them, even trying to follow them up the fields, that they managed to escape only by running and leaving it behind.

38 Christmas in the country

'Christmas in the country,' our suburban friends enthused. 'How marvellous.' They had visions of yule logs, roast goose, mistletoe and happy yokels dancing in the squire's hall. Not on your life! Everyone was grafting away, trying to cash in on the Christmas trade.

Gypsy John arrived in search of holly. I marched him hopefully round the hedgerows and copses but ours did not have sufficient berries to make it saleable. A pity because it was said to be fetching a good price.

He was a slim, dark young man who made his living from buying and selling. He had meticulous manners, was very proud of his Romany ancestry and worried about the disrepute 'newcomers' were bringing to his people. 'Anyone who can get hold of a caravan calls himself a gypsy these days,' he complained. 'They're just drop-outs but people blame us for what they do.'

That same evening big Geoff Bradley, whose land joined Howard's, came into the Forge bemoaning the problems that went with a surfeit of turkeys. 'We've got damn near 500 to get rid of,' he complained. 'It's murder down there. My missus and half-a-dozen others are at it like mad things, feathering away, getting them ready to go.'

'Think of the money,' Howard teased, winking at the rest of us.

'Money?' Geoff echoed as if such a thought had never entered his mind. 'There's a bit, but it's hardly worth the trouble.'

'You giving them away, then?' Howard asked, unrepentant. 'Showing the Christmas spirit?'

'All right, all right, let's talk about something else,' Geoff said and everyone laughed.

The next afternoon we went along to the village school to hear the children singing carols. It was an unpretentious, red-brick Victorian building but a happy place under a spinster headmistress who regarded the forty children as her family and friends and was invited, and came, to their birthday parties and celebrations. The other teacher was a bright young thing in her early twenties who, on one occasion, dealt with a too boisterous Nicholas by upending him in a waste paper bin. He instantly recognized her as a kindred spirit and thereafter thought she was wonderful.

The headmistress played the piano and the young teacher conducted the three rows of scrubbed, Sunday-dressed children. Vicky had a sweet voice and sang one of the solo parts; Nicholas was enthusiastic, had an extremely loud voice and added volume

The young, clear voices rose and fell in traditional tunes and filled the high-ceilinged room. It was a most appreciative audience. Not a chair-scrape, not a sneeze or a cough. Dads with calloused hands, bobbin-bright Mums, brothers and prettied-up older sisters (because you never know who might be there), all sat entranced.

The last item was a modern piece, 'Lord of the Dance'. The children sang it with great verve and gusto and Nicholas Paul and some other boys were carried away enough to stamp their feet, fortunately in time, to add to the sound. It was a touching occasion endowed with the old, pure spirit of Christmas.

Later in the week Shirley arrived back from town with a Christmas tree in the back of the 1800 and John and the kids spent a happy couple of days dressing it to stand in a corner of the sitting room. They went round the farm and managed to find sufficient berried holly to decorate the house.

But, alas, there was no seasonal cheer in the weather.

Griff had trimmed the Forge with holly, paper decorations and a big, golden bunch of mistletoe which was much used. 'Come and spend a couple of hours with us on Christmas Eve,' he invited. 'There's two boys coming down from the hill to play their guitars and sing. It'll be a break and put you in the proper spirit.'

We needed cheering up. These cold days Shirley went round loaded with woollens, slacks, fur-lined boots, Balaclava helmet and gloves.

'By God,' Old Jonathon said admiringly, 'you looks like a knitted Eskimo.'

'I'm taking no chances,' she told him, which led to some ribald advice to me.

It was a good night. There was quite a crowd with the 'townees' from the Saloon coming round to the Public Bar to mix with the locals.

'They should be good for a beer or two,' Old Jonathon told us with twinkling eyes. He had been down with 'flu and did not look his normal, desiccated-gnome self.

The two boys were cousins and looked alike enough to be brothers. They were tall, long-haired, about seventeen or eighteen, with a touch of wildness about them. They played well but sang better. Traditional carols, old music-hall songs learned from their grandparents, and even a couple of pop pieces. It all went down very well and a hat was passed round for them. Most of what they got was silver.

'Better than rearing turkeys, eh ?' Howard needled.

'It comes a damn sight easier,' big Geoff said.

There was a bonus in the form of a mystery man, Jenkin something-or-other, who turned up unexpectedly. I had not seen him before. He avoided company and lived alone, I gathered, in a semi-derelict cottage-holding up the mountain, running a few sheep for a living and with his dogs for company.

Aaron bought him a beer and he stood looking at the company, a tall, shabbily dressed old man, with a savagely hooked nose, a gaunt patriarch's face and the remains of a bush of red hair. A place was made for him at the fire and slowly, as he sat and listened and exchanged the odd word with Aaron and one or two others who bought him beers, and as the warmth and the alcohol thawed him out, he began to relax.

The boys' entertainment was much to his liking; when it ended he suddenly, without rising, began to sing on his own. His voice was cracked with age and disuse, but at one time it had been good. Even now it commanded attention and everyone stopped talking to listen. He faltered as we turned towards him but carried on and sang 'Oh Come All Ye Faithful', with everyone, including the boys, joining in the chorus. At the end everyone clapped.

And then he stood up and recited, all in Welsh, the words pouring out of him so that even we who did not know the language gained something of the sense of it. There was applause when he sat down again but, abruptly, he finished his beer and strode out of the pub without saying goodnight to anyone. Shirley thought he was crying.

Aaron was worried and went outside after him but the old man had gone. 'It's freezing hard,' he reported on his return. 'He must be close on eighty, perhaps more, and he's got seven miles to walk on a night like this.'

He could not tell us much. 'It was an old Welsh poem about

Christmas,' he said in answer to our questions. 'I know it from when I was a child. My grandfather spoke the language although it never done him any good.'

'Who is he, what's his full name?' I asked.

'The Lord alone knows,' he said. 'He's been there years and never told anyone. A proper Welshman, I should say.'

That had to suffice because no one else knew even that much. We talked about the incident as we walked home along the lane, our shoes crunching on the frozen ground.

'Perhaps someone should try and help,' Shirley said.

But I did not get the impression that intruders, however well meaning, would be welcome.

They say in children's tales that if you creep into the cow stalls on Christmas Eve and watch without being seen, as midnight – the hour when Christ was born – strikes you will see the beasts go down to their knees in homage. It may well be true but I cannot vouch for it because long before the hour struck, all my family were sound asleep in bed.

Perhaps it was something in me lingering from the previous night's festivities, but the next morning when I came down to do the early milking, there was a special feeling about the house. The decorated tree stood in the corner of the biggest room, its tinsel and trinkets shimmering in the light of the fire I poked into life. The family's presents were piled on the carpet beneath the loaded branches. Home-made paper chains were draped from the ceiling beams and Shirley had stuck our Christmas cards on the wooden panelling. I wondered how many other men had come down those same stairs in the dawn light of bygone Christmas mornings and looked on similar scenes.

By the time I returned from taking the churns up the lane, the whole family were up and doing the feeding chores. Many hands made light of the work and we were soon inside and sitting down for breakfast. Then, the table cleared, we opened our presents. They reflected our new life. Practical and functional rather than purely decorative. Long woollen socks and heavy, Shirley-knitted, polo-necked sweaters, a tartan pompom beret for Vicky, an extending telescope for Nicholas Paul, and a box of shotgun cartridges for our hunter, John. Shirley got a quilted dressing gown from us all, and I got razor blades, shaving soap and a volume of poems by R. S. Thomas, the Welsh bard who writes about the country and its real people.

Afterwards the kids gave their presents to the animals. A tin of sardines shared between the cats, a tin of dog food divided between the dogs. The recipients were extremely grateful.

Our Christmas dinner was one of our own chickens and a plump duck presented to us by Ellis the Cowman's daughter, locally grown potatoes and Brussels sprouts from our own garden. It might just have been the freshness or the keen weather, but the meal was one of the tastiest I remember. By the time the pudding appeared, it required a special effort to find room for it. Shirley managed, somehow, to ensure that each of the kids found a silver coin in their portion.

John and I did the evening milking and feeding earlier than usual and we finished the day eating Christmas cake and drinking hefty glasses of Egerton elderberry wine.

The day was enlivened by a series of telephoned greetings from our London friends who, judging from the noise and laughter in the background, seemed to be having a high old time. But we would not have changed places with them and my feelings were summed up by Nicholas Paul when he went to bed so tired he could hardly climb the stairs.

'It's been like a real birthday.'

On the last day of December we rose early, did the milking and chores and set out, with the two younger children, for London and a New Year's Eve party. John remained behind to look after the farm and Shirley's mother arrived to look after and cook for him.

Almost before the car stopped outside our friends' house, Shirley was off to keep a hair appointment. I had grown so used to seeing her muffled in Balaclavas and woollens, it was something of a shock to see her in full war paint and party gear. Old Jonathon and his cronies would have been speechless.

The party was excellent and my new hands proved a great attraction, although I rather resented my wife bringing people over to inspect them. They were the colour of old bricks, the backs carried a considerable growth of hair, but they were my own.

Next morning, New Year's Day, we enjoyed the luxury of lying abed until mid-morning before starting back. We arrived in good time to help with the evening tasks. As expected, John had coped happily and scoffed such enormous quantities of food that even his grandmother, despite her experience with four big sons, was impressed.

 39 The ceiling falls in 

It took us a few days to shake off the London living and a deathly cold spell did nothing to help. The lane was like a ski slope. Vans and cars slid about crazily and it needed a long, hard run to take the 1800 up the first steep section. The kids had a fine time riding a little wooden sledge I had knocked together for them but the postman decided not to risk his van and left the mail at the Forge.

No need to cool the milk in the dairy now. It froze in the churns overnight.

Jock the driver laughed and told me, 'Don't you worry, I'm getting a lot of churns like this. They're having to use a steam hose to thaw them out at the depot, but that's their problem, not yours.'

It was advisable to wear gloves to handle the churns. Touch any exposed metal surface with bare hands and your skin immediately began to stick. It was quite a frightening sensation. The possibility of having one hand stuck to a ten-gallon churn of milk did not attract me.

In the midst of all this, one of the cows chose to produce a strapping black bull calf. He was a very handsome fellow who astonished everyone by appearing to be cold-proof. A gale had ripped a hole in the roof of the calving pen and powdered snow drifted down on the baby. It worried him not one jot. He curled contentedly in the thick straw, and thrived.

We had another plus. The piglets had reached marketable size. Griff loaned us a portable scale to weigh them, and as they averaged about six and a half score pounds, which was what we had aimed for, there was nothing to be gained by keeping them any longer.

'They'm eating your profits now,' Howard said when he visited. 'Get 'em away just as quick as you can.' He helped me pick two gilts to keep back for breeding; another two pigs would go to the butcher to be prepared for our own deep freeze.

The following Monday, Shirley and I loaded the first ten into the van to be taken to market. It began as a well-organized operation but ended like something out of an old Keystone Cops routine. The pigs had no intention of going easily. They allowed themselves to be driven into the collecting yard but then they broke through our carefully erected barriers – which were supposed to guide them into the van – and ran in all directions.

The result was that I had first to catch them and then to carry them, one by one, into the van while Shirley guarded the door. As soon as they were picked up they began to squeal and what with this din, plus the fact they weighed only a few pounds less than me, it was a happy moment when the last one was in and the van doors could be locked.

Our market friend, Tall Stan, gave me a hand unloading at the other end. 'Flooding the place with pigs now?' he asked. 'You'll soon be rich.'

The workman helping remembered the lambs. 'You reckon you counted them right this time, boss?'

'Should have,' I told him. 'Ten went in and ten have come out.'

He cast a quick glance at them. 'That's it. No more, no less. You'm getting better at your sums, boss.'

It was a depressed market and they sold at 180 pence a score.

'That's the trouble with pigs,' Tall Stan said. 'They'm up and down like yo-yos. Maybe they'll be a bit better next week, it often happens.'

It did. The second ten sold for 225 pence a score, making the average about 200 pence. It was clear we were not going to make our fortunes in the pork trade, no matter what Stan said.

On the Friday following the sales of the second batch, Dorrie produced another eleven piglets, all healthy, and the next day Dorfie contributed a further ten: so our pig numbers were about back to their previous level. Both farrowings were early morning affairs, which meant a chill midwife stint, but there were no complications, and afterwards both litters lay snugly under the infra-red lamps which were kept on day and night.

It was a relief when we could move them to the follow-on pens and the sows could take over the job of keeping them warm. A coal strike was threatening electricity supplies and there was talk of cuts. They came in February but, fortunately, the local electricity board worked out a good system for informing farmers when their areas would be affected, which made it possible to adapt working routines. Milking by machine at 4.30 a.m. was grisly enough but it was infinitely preferable to having to hand-milk the herd. For the same reason we pushed the evening milking back to 9 p.m.

Even so, the cuts probably cost us a newly born calf. The cow was strangely indifferent to the weak baby and made no attempt to mother or feed it. We gave it a bottle, injected anti-pneumonia vaccine and put it under an infra-red lamp but that night, when it was most needed because of the temperature drop, the lamp went off in a power cut.

I cursed the miners and everyone else I could think of, climbed out of bed at midnight and again at 3 a.m. to tuck hot-water bottles round the calf and wrap it in straw, but it was wasted effort: it died before the kids went to school.

The Hunt collected the carcase in the afternoon and assured us we were not alone. 'It's the cold,' the huntsman said. 'It strikes into their lungs and cuts them down like flowers.'

It was a nasty blow nonetheless: the death of young animals was never pleasant, even if we had grown more hardened to losses since the death of Plain Jane in the early days.

The weather changed and we began to look for the thaw. When it came there was no need for a thermometer. The bathroom ceiling fell in! First it bellied down with the weight of water collected during the night and then it ripped wide open and there was a deluge. Watcr everywhere. Had anyone arrived on the doorstep at that moment and offered a fiver, they could have had the whole farm, lock, stock and barrel.

I shot up the rickety stepladder and into the loft like a monkey up a stick. Too late! My arrival simply aggravated matters. My weight on the joists made them tremble, no more, but it was all that was needed. The whole rotten, water-sodden ceiling came down in one soggy mess with a stomach-lurching sound.

Shirley heard the 'flump' and dashed in to rescue towels, clothes, anything that might get damaged. She looked up at me through the gap where the ceiling had been and asked, 'Is there anything wrong?'

I mastered an urge to throw the spanner at her. 'No. What made you think there might be?'

'You stupid oaf,' she yelled, 'you've smashed in the ceiling.'

Before I could think of a suitable reply to this unfair accusation, she dashed out of the bathroom carrying the things she had collected.

It was only a few minutes' work to tighten the compression joint that had been opened in the freeze, but the damage was done.

And the horror story was not yet complete. The bottom of the sixty-gallon cold water tank turned out to be riddled with holes which did not leak because they were lightly sealed with lime deposits known in the trade as 'Lilies' and protected by a five-inch deposit of sludge which had accumulated over the years.

I called Thomas for help and advice. 'No good leaving it. The tank will have to come out,' he said.

This necessitated cutting away a joist. While we were struggling with this and manoeuvring to get out the sludged tank with the least possible mess, Shirley was busy on the telephone trying to locate a replacement tank.

Having got out the old one, Thomas and I rattled down to town immediately in Old Lil to collect the new one and get plaster board for the ceiling. Further complications arose because the tank had to be cut to take the pipe fittings; we managed it with an assortment of drills and files.

Not pausing for lunch, we raised the tank into position, connected the pipes so that we had a water supply again, and put in a substitute joist. The next day we put in the new ceiling.

Shirley brought back the things she had saved and stood looking at the finished job. I knew what was in her mind.

'It could have happened just the same in London.'

She would not accept this. 'No. There's always someone to help. Nothing is quite the same.'

The danger was that winter had gone on too long for both of us. Suburbia with its centrally heated houses, evenings in front of the telly, and leisurely, social weekends, was beginning to assume a dreamtime quality. Even commuter trains can be remembered as a very sensible way to get about when you are slopping around in freezing mud. Happily, we were at the bottom of the trough. Relief was coming. Even the thaw that brought down the ceiling was an indication of the change creeping over the countryside. There was a new feeling in the air. There was still snow to come but from then on it was soft, slushy stuff which melted within hours.

40 Shepherd to the lambing flock

March, in that district, was another way of spelling lambs. Previously, in our happy urban ignorance, we had accepted that at a certain time every year there would be newborn lambs in the fields our car passed. They were something to point out to the kids. Like spring flowers. Now we learned just how much hard work went into producing the new crop of lambs.

Even so, at the end of it all, a touch of the old magic still lingered. One day there were only the grey, lumpen ewes, the next there were lambs jigging about for all the world as if, like mushrooms, they had sprung from the turf.

Lambing gave everyone concerned a bad time. It meant broken sleep, night walks and anxiety. It left everyone tired and hollow-eyed.

'You'm bad-tempered enough to fight but too tired to pick a quarrel,' was how Aaron put it. But even he admitted, after a lifetime in farming, that if the crop was good and there were not too many losses, it was also an extremely satisfying time. 'You don't have to wait, Jacky. They'm there in front of you.'

Our little flock had come well through the winter. Their fleeces were a hand's depth thick. When we went down with a bucket of meal, they came running and lined up at the feed troughs although their ration was hardly more than a few mouthfuls for each ewe.

In winter conditions the sheep were the most independent creatures on the farm. They foraged over the fields and found enough in the rough grass to live on, but now the ewes were heavy with lamb and needed to be carefully watched. They tired easily under stress and if things went wrong they gave up quickly.

'You keep a good look-out on them,' Old Jonathon advised. 'You'll save a few lambs, you do that.'

He was right. We made a practice of walking round them at least twice a day and it paid off. On one occasion John found a small ewe horribly tangled in brambles in the gulley and cut her free. She was shaky on her legs but unharmed.

The kids were invaluable. They organized their own 'Sheep Patrol' which rivalled Batman in its dramatic style but saved our legs a good deal of walking and was very effective. They came running up one afternoon – everything was done at the double – and saluted before reporting, 'One sheep upside down, sir. Come quickly.'

The ewe in question was lying head down on a slight slope, feet in the air, quite unable to right herself. It needed only a little help from us to get her right way round again and she was soon back grazing. This happened several times as they got really heavy with lamb.

According to the calendar, our first lambs were due about the last week in the month but everyone warned us to 'start looking a few days earlier'. As a precaution we began bringing the flock into the more accessible and better sheltered Calf Field in the evenings and letting them return to the lower fields where there was more feed each morning.

Nearer the expected time John and I began walking round them at night. It was not a pleasant duty. The alarm snarled at midnight and one of us had to crawl out of bed and venture outside. My old duffel coat, which now smelled like a sick goat, served us both; the hood would pull over a Balaclava helmet. The end result had us looking like ghostly monks floating about the place.

On this particular night the fields were still. There was frost in the air and the wind was sharp on my face. In the pale moonlight I could see the sheep lying on the grass. Their eyes shone luminously in the beam of light when I swung the heavy torch round to check on any particular animal. Sometimes a ewe would get up and walk away, but in general my presence was ignored.

It was shivering, unrewarding work until I turned into the far corner of the field where the hedge-line bent crookedly to form a little dingle protected from the wind.

There they were! Our first lambs! They had not long arrived. One was up on trembling legs. The mother, a small, dark-faced Clun, was still cleaning the second. Nothing was required of me. She knew what needed to be done. We stood looking at one another and then, satisfied that I threatened no danger, she went back to the task. For me, it was a moment that justified all the night walks.

Feeling very Biblical, I tucked the lambs under my arms and, followed by the anxious ewe, carried them to the barn where we had built a straw bale reception centre. A hypodermic needle, ready loaded with vaccine which protected lambs against the most common diseases, rested in a box on a ledge. It took only a few minutes to inject both new arrivals, daub them with a spot of red marker fluid and return them to the ewe.

They could spend the night here safe from cold winds and from the foxes which, the locals alleged, snooped around in the hope of snatching a new-born lamb. Once the lambs were dry and active they were considered to be safe: it was the weak, new arrivals that were believed to be in danger.

Old Jonathon had another warning for us. 'You keep them old sows up tight,' he advised. 'When I was a kid, we had an old sow that'd snatch a lamb quicker than any fox and eat it too.'

We took no chances. Dorrie and Dorfie were both confined to barracks.

From now on the lambs began to arrive steadily. There was a high incidence of twins and two sets of triplets among the first comers. Single lambs tend to be carried for the full period, whereas multiples often come early. One lamb from each treble – the smallest – was taken away and put on the bottle. We thought two lambs enough for any ewe to rear. There was a ewe-milk substitute on the market which proved very successful.

It was a hectic time. What with being tired from broken nights and continued alarms and excursions, we walked round in a slight daze, but the kids loved it. They fetched and carried importantly, they petted and fed the bottle lambs, they kept an eye on the ewes and dashed back to tell us when there was a new arrival. They even walked round the field with a bottle of milk ready to give an extra feed to any newcomers that might look frail.

Their enthusiasm was a little wearing at times, but their help was invaluable. We coped as best we could and went round with fingers crossed, hoping that nothing would go wrong, but, of course, inevitably, something did.

My first experience of sheep midwifery came in the bitter hours of a Friday morning. It was a big, cross-bred ewe. She stood alone in a corner of the field, sides heaving, head hanging between her forelegs with exhaustion. She was unable to clear her lamb and the struggle had brought her close to collapse. There was little resistance when I caught her.

It would have been nice to have heard Howard's voice saying the usual, 'I'll tell you what you wants to do . . .' but there was no one around to help. The ewe lay quietly and I wet my right hand with the carbolic soap solution from the bottle in my pocket and felt inside her for the lamb. Instead of the hoped-for head, my fingers found a back leg and a tail: it was arriving backwards and it was horribly twisted.

In spite of the chill, pre-dawn air, there was sweat on my forehead as I worked. No doubt the locals would have laughed at my fumblings, but at last the trapped leg straightened and came free. From that point on, it was fairly easy. Once the feet were clear, the trick was to time your efforts with the ewe's own attempts to expel the lamb. A couple of minutes or so later, it slid clear and lay on the frosted grass. I remembered what should be done and picked it up by the forelegs to give the 'innards' a chance to fall into the proper position. There was no sign of life. Perhaps the struggle had been too much. I hurriedly cleared the mouth and nose and stretched it out for the ewe's attention.

For what seemed an eternity my first attempts as a shepherd appeared unsuccessful but, quite suddenly, the lamb gasped and began to breathe as the ewe nuzzled and licked life and warmth into the young body. There was no further need for me: I was an intruder now. Half and hour later, before going back to bed, I looked at them again. The lamb was up and suckling and the ewe began to lead him away from me.

All this, of course, was routine stuff for the experienced farmer, but even knowing that did nothing to lessen my sense of achievement. Nor did the fact that the scene would probably be repeated half a dozen times before the end of lambing detract from it. Inside the house I poured a stiff slug of Shirley's potato wine to celebrate. It smelled like, and probably tasted like, paint-thinner but the occasion called for some recognition and our 'cellar' offered no alternative.

Inevitably, not even the strictest watch could avert all tragedy. One morning a ewe was lying dead under the hedge with two tiny lambs huddled against her for warmth and comfort. They joined the two 'treble' lambs in a special straw-bale pen.

Another two orphans were acquired when their mother, a big Kerry Hill ewe, died. She produced the twins and appeared to have finished, but the following morning she was moping and obviously sick. There was a third, dead lamb inside her. That was too much for me to tackle, so John and I lifted her into the van and drove over to the vet. During the lambing season he and his assistant were manning a central point to make themselves available to all the local farmers with such emergencies.

'This is about as nasty as any we've run into yet,' the vet said after a preliminary examination. 'The lamb's there but he's dead and twisted like a corkscrew.'

It took some time, but he finally straightened up, cursed his back, and held up the lamb. It was doubled like a safety pin. 'Dead for some time,' he said. 'Nothing we can do about him. I'll give the ewe a couple of shots but you'd better keep her somewhere handy. I don't give much for her chances.'

He was right. The ewe died in the night, despite the urgent buttings of the twins.

Another pair had to be taken away from an old ewe that produced no milk. They sucked hungrily but there was nothing and they grew steadily weaker until the smaller collapsed. An experienced farmer would have spotted their condition long before John came in carrying them both with the ewe following behind.

'Better get something into them as quickly as we can,' he said and sat with them on his knees, trickling milk down their throats. They were too weak to suck. The warm feed rallied them, however, and they survived to take their place with 'Shirley's kindergarten'.

For several days the ewe came to the pen and called for them. They responded at first and she spent the days grazing in the vicinity of the pen, coming up to the straw walls every now and again. But they soon forgot her and transferred their affections to the bottle holders, ignoring her. She hung around another couple of days but then drifted back to the flock. There was no point in keeping her and at the first opportunity she was taken to market and sold for the processed meat trade.

By the end of the first week in April our lambing was all but finished. We had lost another ewe but managed to foster her offspring on a ewe whose own lamb was born dead. Howard had told me all about skinning the dead lamb and putting its coat over the one to be adopted. These two ewes had lambed close together, though, so we simply rubbed the dead lamb over the living one and presented it to the ewe. She hesitated, appeared undeceived and turned away; the pipe-cleaner foundling tottered after her calling; and she turned back, nosed it and decided to adopt it. We left her cleaning the lamb and smuggled her own away.

Our final tally of contributions to the hounds was three dead ewes and four lambs. Our friends considered that we had done 'not too badly', but it was our personal belief that we had done very well indeed. Rather than tempt fate, we decided to follow local precedent and leave counting the lambs until they had been ringed and tailed, but the kids were not so cautious. They informed us that there were seventy-six living lambs, which was very close to the true total but not strictly accurate according to the diary. When we did count tails, we would be very disappointed if there were not seventy-eight.

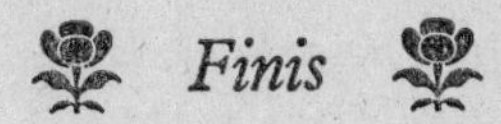

Finis

Lambing completed our first farming cycle. An awful lot of experience had been crammed into it. Since arriving we had planted barley seed, seen it grow, ripen, be harvested and milled into feed. We had made hay and lugged straw. The sows had produced their litters, cows had calved and now we had a crop of lambs. Winter had proved a dreary trial but we had come through without making too many mistakes and had learned a lot.

When we had taken over Egerton, a lifetime ago, spring had been coming in. Now again, the buds were bursting and before long the first leaf would unfurl. And any day at all the new grass would come through.

John Holgate
On a Pig's Back 85p

John Holgate and Family, suburbanites turned farmers, have not only survived their first year on the land . . . they even seem to be getting good at it! Their increasing confidence, however, is not always justified by results as they tackle the muck-spreading and cow-chasing, haymaking and harvest . . . all to the wry amusement of the locals round the bar of the Forge.

'Hilarious' SUNDAY EXPRESS

Edward Vernon
Practice Makes Perfect 85p

Out in the waiting room lurk a confused old lady, a timid vet, a puzzled diabetic, a lonely housewife, a hypochondriac athlete, a tipster with an ulcer, a nun with dandruff, and a persistent young lady with abundant charms and perfect health!

Inside the surgery there is the general practitioner, filling out countless forms, outwitting the pill-freaks, destroying indestructible plastic syringes, watching pharmaceutical salesmen fill his office with gimmickry, and dreading the small-hours phone call from a patient with a hangover!

'An entertaining and often hilarious look behind the surgery door . . . make a bedside book of it' DAILY TELEGRAPH

Patricia Jordan
District Nurse 75p

Born and bred in Belfast, trained in the hard school of student nursing, Patricia Jordan found her niche in a small Northern England town as a district nurse. With all the style of a born storyteller she tells of the patients and their case-histories – the comedy, tragedy and heart-warming humanity of her daily round – and of the doctors and nurses who work alongside her.

'First class . . . admirable reading' OBSERVER

James Herriot

Vet in a Spin 80p

Strapped into the cockpit of a Tiger Moth trainer, James Herriot has swapped his wellingtons and breeches for sheep-skin boots and a baggy flying suit. But the vet-turned-airman is the sort of trainee to terrify flying instructors who've faced the Luftwaffe without flinching. Very soon he's grounded, discharged and back to his old life in the dales around Darrowby.

'Marks the emergence of Herriot as a mature writer'
YORKSHIRE POST

'Just as much fun as its predecessors. May it sell, as usual, in its millions!' THE TIMES

Let Sleeping Vets Lie 80p

The hilarious revelations of James Herriot, the now famous vet in the Yorkshire Dales, continue his happy story of everyday tribulations with unwilling animal patients and their richly diverse owners.

'He can tell a good story against himself, and his pleasure in the beauty of the countryside in which he works is infectious' DAILY TELEGRAPH

Vet in Harness 80p

With the fourth of this superb series James Herriot again takes us on his varied and often hair-raising journeys to still more joyous adventures in the Yorkshire Dales.

'Animal magic . . . James Herriot provokes a chuckle or a lump in your throat in every chapter' DAILY MIRROR

David Taylor
Zoovet 75p

The drowning hippopotamus and the arthritic giraffe . . . The pornographic parrot and the motorcycling chimp . . . Just a few of the patients that are all in a day's work for David Taylor, one of the world's most unusual vets. *Zoovet* is his story of the hilarity and the heartache of animal-doctoring by jetliner across the globe.

'Good humour and abounding energy on every page'
WASHINGTON POST

Molly Weir
Best Foot Forward 75p

'I was engaged as a typist at the breathtaking sum of fifteen shillings a week! A shilling for every year of my age.' In the grim, poverty-stricken world of the Glasgow tenements people met each challenge with pride – and with their best foot forward.

Molly Weir takes up her autobiography where *Shoes Were for Sunday* finished, and as we follow her from her despair at the death of her Grannie to her jubilation at securing her first job, the Glasgow of the time and its salty humour are vividly recreated.